I0008736

Ham Radio Handbook

Baofeng Basics for Beginners

by

Nathan Venture, D

Copyright 2023 Well-Being Publishing. All rights reserved.

No part of this book may be reproduced in any form or by any electronic or mechanical means including information storage and retrieval systems, without permission in writing from the author. The only exception is by a reviewer, who may quote short excerpts in a review.

Although the author and publisher have made every effort to ensure that the information in this book was correct at press time, the author and publisher do not assume and hereby disclaim any liability to any party for any loss, damage, or disruption caused by errors or omissions, whether such errors or omissions result from negligence, accident, or any other cause.

This publication is designed to provide accurate and authoritative information with regard to the subject matter covered. It is sold with the understanding that the publisher is not engaged in rendering professional services. If legal advice or other expert assistance is required, the services of a competent professional should be sought.

The fact that an organization or website is referred to in this work as a citation and/or a potential source of further information does not mean that the author or the publisher endorses the information the organization or website may provide or recommendations it may make.

Please remember that Internet websites listed in this work may have changed or disappeared between when this work was written and when it is read.

To You,

Thank you!

Table of Contents

Chapter 1:
Ham Radio Handbook:
Baofeng Basics for Beginners

Welcome to the world of amateur radio, an exciting and vital field that combines the joy of experimentation with the seriousness of professional communication. Whether you're drawn to ham radio out of curiosity about the airwaves, a desire to connect with others across the globe, or a need to prepare for emergencies, you're starting a journey that can be incredibly rewarding. This chapter aims to lay the groundwork for that journey, focusing on Baofeng radios—a popular entry point for beginners into the amateur radio community.

Baofeng radios have emerged as a cost-effective and accessible option for those new to ham radio. Despite their affordability, these radios pack a punch, offering capabilities that can jumpstart your amateur radio journey. But, before diving deep into the technical aspects, it's crucial to understand what amateur radio is and why it remains relevant in today's digital age.

At its core, amateur radio is about communication and experimentation. It's a hobby, an educational tool, and at times, a lifeline during emergencies. Ham radio operators enjoy communicating with fellow enthusiasts, using radios to explore the technical mysteries of the electromagnetic spectrum. Given its wide appeal, the amateur radio community is diverse, comprising technology enthusiasts, emergency responders, and outdoor adventurers, among others.

Baofeng radios, specifically, offer a gateway into this varied and vibrant world. They are known for their ease of use, versatility, and affordability—key factors that make them ideal for beginners. With a Baofeng radio in hand, users can listen to various communications, including but not limited to, emergency services, the International Space Station, and of course, other ham radio operators.

Embarking on this adventure, the first thing to recognize is the potential of your Baofeng radio. It's not just a communication device; it's a tool for learning and discovery. Its features and capabilities allow beginners to get hands-on experience with radio operation, antenna theories, and electronic fundamentals, all of which are invaluable assets in the ham radio hobby.

Understanding the basics of radio technology and operation is the first significant step. For instance, knowing the difference between UHF and VHF frequencies, or how to program your Baofeng radio to access local repeaters, can drastically enhance your experience. These technical details form the foundation upon which you can build your ham radio expertise.

Additionally, amateur radio is steeped in etiquette and protocol, much of which has been developed over decades of operation. Familiarity with these conventions will not only enrich your understanding but also help you integrate into the ham radio community more smoothly. As you progress, you'll find that this community is one of the greatest resources available—a group eager to share knowledge, provide assistance, and inspire others.

Moreover, getting started with a Baofeng radio is exceptionally practical. The barriers to entry, in terms of cost and complexity, are relatively low. This accessibility empowers more people to explore amateur radio, ensuring a continual influx of fresh enthusiasm and ideas into the hobby. It's a dynamic environment where learning is ongoing, discoveries are waiting, and friendships span the globe.

It's worth noting that, while the technical aspect of ham radio is fascinating, the real magic lies in its ability to connect people. Through the airwaves, you can touch the lives of individuals in far-flung corners of the world, share experiences with fellow enthusiasts, or provide critical communication during emergencies. This aspect of amateur radio not only adds a human dimension to the hobby but also underscores its significance beyond mere technology.

As you venture into the world of Baofeng radios and amateur radio at large, remember that patience and perseverance are your allies. The learning curve might seem steep at first, but each step forward brings you closer to mastering this captivating form of communication. And with each new skill acquired, you'll find yourself more deeply connected to a community that's both welcoming and inspiring.

So, as we move forward, consider this chapter your first step into a larger world. We'll cover the essentials of choosing and using a Baofeng radio, from setting up your device to making your first contact. We'll demystify the technical jargon and provide practical guidance to help you become an active and confident ham radio operator.

Remember, your journey in amateur radio and with your Baofeng begins with curiosity and evolves through learning and practice. It's about embracing challenges, celebrating successes, and always reaching for that next level of achievement. Welcome aboard. The airwaves await, filled with opportunities for adventure, learning, and connection. Let's get started!

Through the chapters that follow, you'll gain insight into the broader aspects of ham radio, delve into the specific features and capabilities of Baofeng radios, and equip yourself with the knowledge needed to thrive in this exciting hobby. From understanding radio fundamentals to making your first contact and beyond, we're here to guide you every step of the way.

The goal is not just to educate but to inspire and motivate you to explore, connect, and grow within the amateur radio community. So, whether you're looking at ham radio as a hobby, a means of emergency preparedness, or a gateway to outdoor adventures, you're embarking on a journey that promises to be as enriching as it is enlightening. Welcome to the world of Baofeng and ham radio. Let the adventure begin!

Introduction

Welcome to the world of amateur radio, a fascinating journey that will take you from the basics of Baofeng radios to mastering the art of radio communication. Whether you're a hobbyist, a prepper, an outdoor adventurer, or simply curious about the technology, this book is designed to guide you through the intricacies of ham radio. Our aim is to provide you with a comprehensive understanding that not only enlightens but also empowers you to harness the full potential of amateur radio for your personal or professional growth.

At the heart of every adventure is the thrill of discovery and learning. Amateur radio offers an unmatched opportunity to explore a world where communication transcends the boundaries of geography, allows for emergency preparedness, and fosters a global community of enthusiasts. Imagine being able to connect with people across the globe, share ideas, and learn from others, all through the power of radio waves. That's the promise of amateur radio, and it's more accessible now than ever before, especially with tools like Baofeng radios.

For those new to the scene, the concept of radio communication might seem daunting at first glance. However, like any skill worth acquiring, the journey from novice to proficient is one of gradual, rewarding steps. This book is tailored to make that transition as smooth and enjoyable as possible, armed with practical guidance and insights that bridge theory with real-world application.

Why Baofeng, you might wonder? Baofeng radios have revolutionized the way we think about and engage with amateur radio. Known for their affordability, versatility, and ease of use, these radios serve as an excellent starting point for anyone eager to dive into the world of radio communication. They embody the idea that access to knowledge and tools can transform hobbyists into skilled communicators, capable of navigating both the joyous and challenging aspects of amateur radio.

Embarking on this journey requires more than just technical know-how; it demands curiosity, patience, and a thirst for knowledge. This book aims to kindle that curiosity, providing a mixture of descriptive, instructional, motivational, and inspirational content that guides you through the basics of radio operation, setup, and etiquette, all the way to advanced techniques and emergency preparedness.

Understanding the significance of amateur radio in today's world is crucial. Far from being a relic of the past, amateur radio is a dynamic field that continually evolves with advancements in technology. It plays a pivotal role in emergency communication, community building, and technological experimentation. By mastering the art of radio communication, you're not just engaging in a fulfilling hobby but also equipping yourself with vital skills that could one day prove invaluable in crisis situations.

As you delve into the chapters of this book, you will find a structured path laid out before you. Starting with the foundational aspects of ham radio and the unique benefits of Baofeng radios, you will gradually move towards understanding the technicalities of radio communication, setting up your equipment, and making your first contacts. Each section is crafted to build upon the last, ensuring a cohesive learning experience that respects your pace and curiosity.

The journey towards becoming an adept member of the amateur radio community is as much about the technical skills acquired as it is about the connections made and the shared spirit of exploration. Through this book, we aim to foster a sense of belonging and enthusiasm within the ham radio community, drawing on the collective wisdom and experiences of those who have journeyed before you.

Radio communication is a doorway to endless possibilities. It's a bridge between people, ideas, and cultures. As you progress, you'll discover not only the sheer joy of making your first contact but also the satisfaction that comes from mastering a technology that can connect, inform, and sometimes even save lives. This is a realm where the old and the new merge, where digital and analog technologies coexist, and where every dial turned and button pressed adds to the tapestry of global communication.

In the chapters that follow, practical advice is interwoven with inspirational insights, designed to motivate you to explore, experiment, and expand your horizons. We traverse from the essentials of navigating your Baofeng radio to the rich tapestry of the amateur radio community, highlighting how you can contribute and grow within this vibrant ecosystem.

The journey of amateur radio is one that is richly rewarding, offering a unique blend of technical challenge, personal accomplishment, and community connection. It's about more than just transmitting messages; it's about becoming part of a global family of enthusiasts who share your passion for communication and technology.

Through this book, we aim not only to educate but also to inspire you to reach out into the ether and make connections that span continents. We hope to embolden you with the knowledge and skills necessary to not only navigate the world of amateur radio but to thrive within it, embracing its challenges and opportunities with confidence and curiosity.

Let this journey transform not just your understanding of radio communication, but also your perspective on what it means to connect in an increasingly digital world. The path from beginner to enthusiast is a journey worth taking, filled with learning, growth, and a myriad of possibilities. Welcome to the exciting world of amateur radio – your adventure starts here.

Chapter 2:
Understanding Ham Radio

Diving deeper into the world of ham radio opens up a universe where the past and the future converge through waves and frequencies. Ham radio, or amateur radio, isn't just a hobby; it's a lifeline connecting diverse individuals across the globe, serving roles from emergency communication to leisurely chats between distant lands. Its roots stretch back to the early days of radio, showcasing a history rich with innovation, community, and a relentless pursuit of connecting voices without the need for wires or the internet. Today, ham radio remains relevant by adapting to modern needs while holding onto the essence of direct, person-to-person communication. This chapter navigates through the exhilarating evolution of amateur radio, showcasing its significance not just as a technical hobby but as a vital tool for global connectivity. The myriad ways in which ham radio plugs into the fabric of current communications, defying the onslaught of instant messaging and social media, is nothing short of inspiring. As you immerse in the history and the undeniable relevance of ham radio today, you're stepping into a world where every broadcast and every connection tells a story, inviting you to be part of a continuum that is both a nod to the pioneers of radio and a handshake with the infinite possibilities of the future.

History and Evolution of Amateur Radio

The journey of amateur radio, commonly known as ham radio, is a testament to human ingenuity and the relentless pursuit of communication. From its early days, this fascinating world has evolved, adapting and advancing with every technological break-through. It's a narrative not just about technology, but about society, community, and the unyielding human desire to connect and communicate.

The earliest beginnings of amateur radio can be traced back to the late 19th century, with pioneers like Guglielmo Marconi experimenting with wireless telegraphy. These early tinkerers and inventors laid the groundwork for what would become a global community of enthusiasts, pushing the boundaries of what was possible with radio waves.

By the early 20th century, amateur radio operators were not only experimenting with technology but were also providing vital communication during emergencies. Their ability to send messages across long distances, without the need for a physical connection, proved invaluable in disasters when traditional lines of communication were down.

However, the growth of amateur radio was not without its challenges. The spectrum is a finite resource, and as commercial and military interests in radio technology grew, amateurs found themselves competing for space on the airwaves. Regulatory bodies began to emerge, tasked with arbitrating these competing interests and ensuring that the spectrum was used efficiently and responsibly.

In the United States, the Federal Communications Commission (FCC) was established in 1934, marking a new era of amateur radio. With this formal recognition came rules and regulations, including the requirement for operators to obtain licenses. These regulations not only legitimized the hobby but also helped to organize and standardize practices, ensuring amateurs could continue to innovate while coexisting with commercial and military users.

The mid-20th century saw amateur radio operators increasingly contributing to scientific research and emergency response. Hams provided essential communication during World War II, further demonstrating the value of amateur radio to national security and public safety.

As technology advanced, so did amateur radio. Transistors replaced vacuum tubes, making radios smaller, more affordable, and accessible to a broader audience. This democratization of technology sparked a surge in interest, with clubs and organizations sprouting up around the world to support the growing community.

The advent of the internet in the late 20th century presented both a challenge and an opportunity for amateur radio. While some feared that digital communication would render amateur radio obsolete, the community found innovative ways to integrate new technologies, embracing digital modes of communication that expanded the hobby's capabilities and appeal.

Today, amateur radio is a vibrant and diverse hobby, with over three million licensees worldwide. It encompasses a wide range of activities, from traditional voice communication to digital modes, satellite communication, and even radio astronomy. The community's spirit of experimentation and innovation remains as strong as ever, driven by a passion for exploration and the joy of connecting with others.

One of the most compelling aspects of amateur radio is its role in emergency preparedness and response. Hams are often among the first to provide critical communication services following natural disasters, using their skills and equipment to bridge the gap until traditional services can be restored. This commitment to public service exemplifies the ham spirit — a combination of technical proficiency, ingenuity, and a desire to help others.

Amateur radio has also played a crucial role in education, inspiring generations of young people to pursue careers in science, technology, engineering, and math (STEM) fields. Many hams credit their early experiences with amateur radio as the spark that ignited their passion for science and technology.

As we look to the future, the potential of amateur radio is boundless. Emerging technologies like software-defined radio (SDR) and low-earth-orbit (LEO) satellites offer exciting new frontiers for exploration. The community's relentless pursuit of innovation promises to keep amateur radio at the cutting edge of technology, while its spirit of camaraderie and public service continues to make it a deeply rewarding hobby.

For those just beginning their journey into the world of amateur radio, the history of this remarkable hobby is a source of inspiration. It's a story of persistence, creativity, and community that underscores the transformative power of communication. As we turn each new page in the evolution of amateur radio, we're reminded that at its heart, this hobby is about more than just radios — it's about connecting people.

Joining the amateur radio community means becoming part of this rich history. It means contributing your voice to a chorus that spans the globe, bound together by a shared love of communication and a common spirit of discovery. Whether you're intrigued by the technology, drawn to the opportunity for public service, or simply looking for a unique way to connect with people around the world, amateur radio offers something for everyone.

So, as you embark on this journey, remember that you're not just learning about circuits, antennas, and regulations — you're becoming part of a story that's as old as radio itself. A story that, with every new voice, becomes even more vibrant and extraordinary.

The Role of Ham Radio in Modern Communication

Many might think of ham radio as a relic of the past, a hobby clinging to the fringes of technology. Yet, this perception couldn't be further from the truth. In the current digital age, where global communication is at our fingertips, ham radio serves a distinct and vital role. It's a bridge between old and new, connecting individuals across countries and continents without the need for a cellular signal or internet connection.

Amateur radio operators, or hams, are known for their spirit of adventure and insatiable curiosity. They explore frequencies to make contact with fellow enthusiasts worldwide, experiment with antennas, and even bounce signals off the moon. But beyond the hobbyist aspect, ham radio carries significant weight in modern communication, especially in times of emergency.

When natural disasters strike and conventional communication infrastructure fails, ham radio operators often step in to fill the void. They provide critical links for rescue operations, coordinating efforts and connecting victims with emergency services. This role has been demonstrated repeatedly, from hurricanes to earthquakes, where cellphone towers and internet services were disrupted.

Furthermore, amateur radio is at the forefront of technological innovation. Hams were among the first to experiment with digital communication modes and satellite communication, laying the groundwork for developments in commercial and military communications. Their contributions have not only enriched the hobby but have also advanced the field of communication technology.

In the realm of education, ham radio offers an engaging platform for teaching electronics, communication theory, and radio wave propagation. Schools and universities around the globe incorporate amateur radio into their STEM curriculum, empowering students with hands-on learning experiences that stimulate interest in science and technology careers.

For outdoor adventurers and explorers, ham radio provides a reliable means of communication in remote areas where cell phones dare not roam. It's not just about making contacts; it's about safety, navigation, and the joy of sharing one's adventures over the airwaves. Whether hiking in the backcountry or sailing the open seas, a ham radio can be a lifeline to the outside world.

The community aspect of ham radio also stands out. It's a global network of friends you haven't met yet, where camaraderie and mutual support are the order of the day. Annual events like Field Day and the DX contests unite hams in friendly competition and cooperative endeavors, fostering a sense of global kinship.

Amidst the rapid pace of technological change, ham radio offers a unique blend of tradition and innovation. It's a hobby that requires mastering the art of radio communication and the science behind it. But more importantly, it instills a deep appreciation for the magic of human connection across vast distances.

In terms of emergency preparedness, the relevance of ham radio cannot be overstated. Operators are often among the first to provide eyewitness accounts of events unfolding in real-time, thanks to their independent networks. Their ability to mobilize quickly and create ad-hoc networks is invaluable to emergency responders and relief organizations.

From a regulatory perspective, the ham radio community plays a critical role in advocating for the preservation of radio frequencies for amateur use. Their active participation in international conferences and regulatory bodies helps to ensure that the spectrum remains accessible for educational, technical, and emergency communication purposes.

With the advent of digital modes and software-defined radio, ham radio is experiencing a renaissance. These technologies enable more efficient use of the spectrum, innovative communication techniques, and the integration of computers and radios in ways that were unimaginable a few decades ago.

Ham radio also serves as a testing ground for experimental communication methods. Hams have the freedom to tinker and innovate, leading to breakthroughs that benefit not only the amateur radio community but also commercial and military applications. This spirit of experimentation keeps the hobby vibrant and continually evolving.

Moreover, ham radio acts as a beacon for promoting international goodwill. By fostering direct communication between individuals around the world, it breaks down barriers and cultivates a deeper understanding among different cultures and societies. In times of conflict or natural catastrophe, these connections can prove to be lifelines of hope and solidarity.

In the world of broadcasting, ham radio enthusiasts have pushed the boundaries of what's possible with limited resources. They've launched amateur radio satellites, known as OSCARs, into space, allowing for global communication on a shoestring budget. These achievements showcase the ingenuity and resourcefulness of the ham radio community.

In conclusion, ham radio remains a significant player in the modern communication landscape. It blends passion with purpose, serving not just as a pastime, but as a vital tool for education, innovation, emergency response, and fostering global connections. As we look to the future, amateur radio continues to adapt and thrive, proving that this time-honored hobby holds enduring relevance in our increasingly connected world.

Chapter 3:
Getting Licensed

Embarking on the journey to becoming a licensed ham radio operator is a pivotal step in unlocking a world of communication possibilities. It's more than just acquiring a legal permit; it's about embracing a community and mastering a skill set that can be life-changing. The process might seem daunting at first glance, but with the right approach, it's entirely achievable. To start, familiarize yourself with the various license classes—each offering different privileges and opportunities on the airwaves. Diving into study materials specifically designed for the ham radio exam will build a strong foundation of knowledge. Remember, understanding is key; it's not just about memorizing answers for the test. There are a plethora of resources available ranging from books to online courses and practice exams that cater to everyone's learning style. Taking the exam itself is a milestone, and finding a comfortable, supportive environment to do so can greatly ease any nerves. As you prepare, keep in mind that every seasoned ham operator started exactly where you are now. They, too, navigated the path of studying, understanding regulations, and practical operating principles. Your determination to succeed and the support of the ham radio community will guide you through. Once licensed, a vast world filled with the art and science of radio communication awaits. This chapter is designed to not only prepare you for getting your license but to inspire a lifelong journey in the fascinating world of amateur radio.

Study Tips for the Ham Radio Exam

Embarking on the journey to become a licensed ham radio operator is a thrilling adventure, a blend of modern and historical, technical and personal. Picture this - a community where every member extends a hand, where the airwaves unite enthusiasts from all corners of the globe, bridging isolation with a network of voices. That is ham radio. But before you can join this vibrant community, you need to ace that exam. Let's dive into some powerful study tips that can help you succeed and start your journey in ham radio.

Firstly, understand that the journey to passing the ham radio exam starts with a simple step – familiarizing yourself with the basics. The exam covers a range of topics including regulations, operating practices, and electronics theory. Breaking down these topics into manageable sections can prevent you from feeling overwhelmed. A steady pace wins this race. Dedicate time every day to focus on one aspect of the exam. Gradually, you'll build a robust foundation that will serve you not only in the exam but in your ham radio journey.

Resources are abundant. Make use of study guides, online tutorials, and practice exams. These tools are invaluable as they offer insights into the format of the actual exam and the types of questions you can expect. A variety of study materials are beneficial to cater to different learning styles, whether that's visual, auditory, or kinesthetic. Don't hesitate to mix and match these resources to suit your unique learning style.

Joining a study group can be a game-changer. There's strength in numbers. Engaging with fellow enthusiasts who are also preparing for the exam can motivate you and clarify complex concepts. Study groups encourage accountability, a crucial component of effective study habits. Sharing knowledge and discussing topics not only reinforces your learning but also embeds the information deeper into your memory. So reach out, connect, and grow together.

Practice exams are more than just a test of knowledge; they are a window into the exam mindset. Making them a regular part of your study routine can highlight areas that need more focus while building your confidence. It's one thing to understand a concept theoretically; it's another to apply it under exam conditions. Practice exams are your rehearsal for the real performance.

One of the keys to retaining information is to teach it. Explaining concepts to a friend or family member can unveil a new depth of understanding for both you and your audience. This strategy not only reinforces your own learning but also helps in identifying any gaps in your knowledge. Whether it's the intricacies of antenna theory or the specifics of FCC rules, teaching is learning twice.

Setting tangible goals can significantly enhance your study experience. These can range from dedicating a specific amount of time each day for study, to mastering a particular topic by the end of the week. Goals give direction and purpose to your preparation, making your study sessions more focused and efficient.

Remember, it's essential to give yourself breaks. Marathon study sessions can lead to burnout. The brain assimilates and processes information more effectively with regular, well-timed breaks. Your study time should be sustainable; think of it as a marathon, not a sprint. Incorporating breaks into your study routine can refresh your mind, making it ready for the next round of learning.

Visualization can play a critical role in your preparation. Visualize not just success in the exam, but the practical application of the knowledge you're acquiring. Imagine setting up your own ham radio station, contacting other operators worldwide, and joining a community that shares your passion. This vision can be a powerful motivator, driving you forwards even when the material becomes challenging.

Feedback is your friend. As you practice and study, seek feedback on your progress. This could be from your study group, online forums, or even practice exam results. Constructive feedback can guide your study direction, sharpening your focus on areas that require more attention.

A healthy body harbors a healthy mind. Don't neglect your physical well-being in the quest to excel in the exam. Regular exercise, adequate sleep, and proper nutrition can significantly impact your cognitive functions, enhancing memory retention, focus, and overall mental clarity.

On the day before the exam, resist the urge to cram. Instead, review your notes lightly, relax, and get a good night's sleep. Approaching the exam with a calm and rested mind can make a significant difference in your performance.

Lastly, approach the exam with confidence. You've prepared, practiced, and now it's time to trust in your capabilities. A positive mindset can alleviate anxiety and enhance your ability to recall information. Remember, this exam is not just a hurdle; it's a gateway to a world of opportunity in ham radio.

In the end, passing the ham radio exam opens up a universe of possibilities. It's not just about getting licensed; it's about becoming a part of a global community that shares your enthusiasm for communication, technology, and connection. With the right preparation, strategies, and mindset, you can embark on this exciting journey with assurance and excitement. The airwaves await, ready to carry your voice across the globe. Your adventure in ham radio starts now.

Embrace the challenge, immerse yourself in the learning process, and let your passion for ham radio guide you. The journey ahead is not just about mastering the technicalities of radio communication; it's about joining a fellowship of enthusiasts who are ever ready to explore, discover, and innovate. Welcome to the world of ham radio, where every frequency holds a story, and every call sign marks the beginning of a new friendship. Let's turn the dial, tune into the future, and embark on this mesmerizing voyage together.

Understanding the Different License Classes

Embarking on the journey of becoming a licensed ham radio operator is both thrilling and a bit daunting. The world of amateur radio is vast, filled with potential for exploration, connection, and lifelong learning. A pivotal step in this journey is understanding the different classes of licenses available. Each class not only signifies a level of proficiency and commitment but also opens new avenues for communication and experimentation.

In the realm of amateur radio, there are three primary license classes in the United States: Technician, General, and Amateur Extra. Each class offers its unique set of operating privileges across the amateur radio bands—vast frequencies allocated specifically for amateur use. Understanding these classes is crucial because the license you hold determines not just how you can communicate but also what realms of the radio spectrum you can explore.

The Technician license, often seen as the entry point into ham radio, grants access to all amateur radio frequencies above 30 megahertz. This includes the very popular 2-meter band. Technicians can also operate on certain high-frequency (HF) bands using Morse code. This class is designed to get you started, offering a taste of what ham radio has to offer while laying a foundation in radio theory and operating principles.

General class license holders enjoy significantly expanded privileges. In addition to the VHF and UHF frequencies available to

Technician licensees, General class operators gain access to portions of nearly all HF bands. This opens up a world of global communication, allowing for direct contact with other amateur radio enthusiasts across the globe. The General license is a gateway to the heart of what many consider the true essence of ham radio.

At the pinnacle is the Amateur Extra license. This class offers unrestricted access to all amateur frequencies. Achieving an Amateur Extra license signifies a deep understanding of radio theory, electronics, and FCC regulations. It's a commitment to excellence and a passport to elite segments of the amateur bands which are reserved exclusively for operators who have passed this most rigorous examination.

The process of obtaining these licenses involves passing written examinations that progressively increase in difficulty. The Technician exam covers basic regulations, operating practices, and electronics theory. The General and Amateur Extra exams delve deeper into these topics, challenging candidates to demonstrate a thorough under-standing of more complex concepts and regulations.

Why does this matter? The answer lies in the essence of amateur radio itself. It's not just about communication; it's about exploration, education, and contributing to a global community. Each step up the licensing ladder not only broadens your operating privileges but also deepens your engagement with the hobby. It encourages continuous learning and improvement, traits that epitomize the spirit of ham radio.

As you consider which license class to pursue, think about your interests and goals in amateur radio. If you're excited by the prospect of chatting with locals on a repeater, the Technician license may fulfill your needs. But if you dream of bouncing signals off the ionosphere to make contact with someone halfway around the world, you'll likely find the General or Amateur Extra class more appealing.

Remember, obtaining your license is not just about passing an exam; it's about joining a community. Amateur radio operators are known for their willingness to mentor newcomers. Many find that preparing for their next license class is an exciting journey of personal growth, facilitated by the support and encouragement of the ham community.

Moreover, the advancements in technology and the introduction of new digital modes have made the world of amateur radio more accessible and rewarding than ever before. Whether you're interested in traditional voice communication, digital modes, or even satellite operation, there's a place for you in amateur radio.

As you embark on this journey, let the prospect of what lies ahead fuel your enthusiasm. Imagine the thrill of your first contact, the satisfaction of mastering a new mode of communication, or the pride in contributing to a community service project. These experiences and many more await you in the world of amateur radio.

In conclusion, understanding the different license classes is crucial for any aspiring amateur radio operator. Each class offers unique privileges and opportunities, and choosing the right path is a personal decision that depends on your interests and goals. Whether you're drawn to the hobby for the joy of experimentation, the thrill of competition, or the satisfaction of public service, there's a place for you in the amateur radio community. Embrace the journey, and let it inspire you to new heights of achievement and camaraderie. Welcome to the endlessly fascinating world of amateur radio.

How to Take Your Exam

Embarking on the journey to become a licensed ham radio operator is an exciting step toward joining a global community of passionate communication enthusiasts. The process involves dedication, study, and passing your exam, which is a pivotal moment in your ham radio journey. This section will guide you through preparing, scheduling, and taking your amateur radio license exam, setting you on the path to endless exploration and connection.

The first step in your adventure is to understand the structure of the exam. Depending on the license class you are aiming for - Technician, General, or Extra - the complexity and content will differ. However, each exam consists of multiple-choice questions drawn from a publicly available pool, covering a range of topics essential to operating successfully and legally on the airwaves.

To register for your exam, you will need to locate a Volunteer Examiner Coordinator (VEC) session in your area. The American Radio Relay League (ARRL) website, among others, provides a searchable database of upcoming exam sessions. Registration details vary, but it's common to need to sign up in advance and sometimes pay a small fee.

Preparation is key to success. Beyond understanding the theory, immersing yourself in the practical aspects of radio operation can vastly improve your grasp of the material. Joining a local ham radio club can provide invaluable hands-on experience and mentorship from experienced operators who are usually more than willing to support newcomers.

On exam day, ensure you arrive with everything you need. This typically includes valid identification, your Social Security Number (SSN) or Federal Registration Number (FRN), a calculator (if permitted), and the exam fee. Check with your VEC for specific requirements ahead of time to avoid any surprises.

During the exam, time management is crucial. While the questions are multiple choice, some require calculations or application of concepts that can be time-consuming. Skim through the exam to answer questions you're confident about first, then return to tackle the more challenging ones.

Remember, this exam is not just a hurdle; it's a way to ensure you have the necessary knowledge to enjoy and participate in amateur radio safely and effectively. Approach each question with confidence and remember that thorough preparation will shine through in your answers.

If you don't pass on your first attempt, don't be discouraged. Many don't. Use it as a learning experience. Review which areas were challenging and focus your studies on those topics before retaking the exam. The vast majority of people improve on their second attempt.

Upon passing the exam, you will receive a CSCE (Certificate of Successful Completion of Examination) document. This is your proof of passing until your official license arrives. The VEC will submit your results to the FCC, after which you can typically expect your new call sign to appear in the FCC's database within a few days.

With your call sign in hand, you are now part of the vibrant global community of amateur radio operators. This license is not just a certificate; it's a passport to a world of exploration, learning, and camaraderie. You've earned the right to explore frequencies, make connections across the globe, and contribute your voice to the airwaves.

Remember, the journey doesn't stop at getting licensed. The world of amateur radio is vast and varied, offering continuous opportunities for growth and exploration. Whether your interest lies in emergency communication, digital modes, contesting, or simply chatting with fellow ham operators worldwide, your license is the beginning of what can be a lifelong journey.

As you set out on this new adventure, stay curious and open-minded. Amateur radio is a hobby where learning never stops, and each day can bring something new and exciting. Engage with your local ham community, participate in on-air events, and never stop expanding your knowledge and skills.

Finally, remember the significance of what you've achieved. Passing your amateur radio exam and receiving your license puts you among a select group of individuals who've taken the time and effort to contribute positively to the worldwide community of communications enthusiasts. You're now equipped not just with a license, but with the ability to make real impacts in times of emergency, to advance the science of radio, and to bring people together, no matter where they are.

Welcome to the world of amateur radio. It's a community that thrives on the shared passion for communication, innovation, and connection. As you embark on your journey of exploration and discovery, know that you are a vital part of this exciting and continually evolving field. Here's to the many adventures and learning experiences that await you on the airwaves!

Chapter 4:
The World of Baofeng Radios

In threading through the vibrant tapestry of amateur radio, we venture into a realm particularly beloved by newcomers and seasoned enthusiasts alike: the world of Baofeng radios. This chapter unveils the allure that wraps around these compact, budget-friendly devices, casting a light on why they stand as a beacon for those embarking on their ham radio journey. Baofeng handheld transceivers, with their array of models and features, present a gateway to radio communication, inviting curiosity and demystifying the airwaves for the uninitiated. They aren't just tools; they're entry tickets to a global dialogue, where voices from across continents converge. Here, we explore how Baofeng radios, despite their modest pricing, are packed with capabilities that can rival their higher-end counterparts, making them an exemplary choice for beginners. This isn't about settling for less; it's about smart starts and strategic learning. Through the lens of Baofeng, we delve into the essence of amateur radio—communication unwired, unbound, and utterly fascinating. As you turn the pages of this chapter, let it be a beacon guiding you towards unlocking the potential that lies in the palm of your hand, resonating with the broader narrative that anyone, with just a spark of interest and the right tools, can master the art of radio communication.

Introduction to Baofeng Handheld Transceivers

Embarking on the journey of ham radio communication opens up a universe of endless possibilities, where every turn presents an opportunity to connect, learn, and thrive in a community bound by the airwaves. At the heart of this adventure lies the Baofeng handheld transceiver, a gateway for enthusiasts venturing into the fascinating world of amateur radio. Baofeng radios stand out as the Swiss Army knife in the communications toolkit of hobbyists, preppers, and outdoor adventurers alike, offering a blend of affordability, accessibility, and versatility unmatched by many. As we dive into the intricacies of radio communication, understanding the features, operation, and potential of these devices becomes paramount. Whether it's for emergency preparedness, connecting with fellow hobbyists, or simply exploring the natural wonders through radio frequencies, Baofeng transceivers serve as an essential companion, empowering users with the means to embark on their ham radio journey. These devices not only demystify the complexities associated with radio communication but also lay the foundation for a rewarding experience that transcends the conventional boundaries of interaction. As we unfold the pages of this exploration, remember, the journey with Baofeng is not just about mastering the technicalities; it's about embracing the vast, vibrant community and the boundless adventures that amateur radio has to offer.

Model Comparisons and Features

As we delve further into the world of Baofeng radios, an essential step for any aspiring radio enthusiast is understanding the different models and their unique features. Each model under the Baofeng name comes with its set of capabilities, designed to cater to various needs and preferences - from the rugged outdoor adventurer to the detail-oriented hobbyist.

At the core, Baofeng radios are known for their affordability and reliability, making them an excellent entry point for beginners in amateur radio. However, with such a wide range on offer, it's vital to know what sets each model apart to make an informed decision that aligns with your needs and aspirations in the radio communication world.

Take, for instance, the UV-5R, one of the most popular models among beginners. Its appeal lies in its balance between features, durability, and price. It offers dual-band capabilities, allowing users to explore both VHF and UHF frequencies. Its ease of use combined with the essential features needed for clear communication makes it a go-to choice for those just starting their journey.

On the other hand, the BF-F8HP stands out for its high power output, leapfrogging over the UV-5R, with up to 8 watts of power. This particular feature significantly improves the range and clarity of transmissions, which can be a game-changer in various situations, especially for outdoor adventures or emergency preparedness where communication distance matters.

Comparatively, models like the GT-3WP have carved a niche for themselves with their waterproof design. Given the unpredictable nature of outdoor environments, the GT-3WP provides peace of mind through its durability and resistance to water, making it an ideal companion for those exposed to the elements or involved in water-related activities.

Each Baofeng model also offers a range of accessories and customization options, enhancing their functionality. From extended batteries, improved antennas to programming cables, these extras allow users to tailor their radios to specific needs, thereby enhancing the overall experience and utility of the device.

Programming capabilities further differentiate the models. While all Baofeng radios can be programmed manually, some models are more conducive to computer-based programming, offering a smoother, more streamlined setup for those who prefer or require customized channel lists.

The decision between models may also be influenced by the radio's interface and display. Some users might prefer the simplicity and ease of use found in models with more basic displays, while others might seek out those with more sophisticated, detailed screens that provide additional information at a glance.

Then there's the question of accessories. The compatibility and availability of accessories can significantly affect your choice. For instance, a model that supports a wide range of affordable, easily obtainable accessories could be more appealing than one that doesn't, purely for the convenience and customization options it offers.

Battery life is another critical feature that varies across Baofeng models. For users who require their radio to last longer periods between charges, models boasting enhanced battery life will be more attractive. This feature is particularly important for those relying on their radio in situations where recharging may not be feasible.

The integration of new technology also sets certain Baofeng models apart. Advanced features such as GPS capabilities, which can be significant for navigation and location tracking, are becoming more sought after. While not all models offer this, those that do stand out for adventurers and professionals who benefit from geo-location features.

For those interested in digital modes of communication, certain Baofeng radios offer compatibility with digital mobile radio (DMR) standards, opening up a world of digital communication possibilities.

Although not all models support DMR, the ones that do provide a bridge for users looking to explore digital transmissions.

Audio quality is another area where Baofeng models differ. Even though all are designed to offer clear communication, some models are equipped with enhanced noise reduction technologies and superior speakers, making them preferable in noisy environments or for users with higher audio quality demands.

When choosing your Baofeng radio, it's essential to consider future growth and learning in the amateur radio hobby. A model that might seem fitting for your current needs could become limiting as you grow more experienced and your interests in radio communication expand.

Lastly, community and support are invaluable assets. Some Baofeng models boast a more extensive user base, resulting in more resources like guides, tutorials, and community forums. These resources can be incredibly helpful, providing support and enhancing your learning and troubleshooting capabilities.

In sum, each Baofeng model offers a world of possibilities, tailored to different needs, preferences, and aspirations. Whether you're stepping into the amateur radio community for the first time or looking to expand your horizons within this fascinating hobby, understanding the landscape of Baofeng radios is your first step towards making informed choices that not only satisfy your immediate needs but also fuel your long-term passion and involvement in the world of amateur radio.

Why Baofeng Is a Great Option for Beginners

Embarking on the journey of amateur radio communication often begins with choosing the right equipment. For many beginners, the myriad of options available can be overwhelming. However, Baofeng radios stand out as an excellent starting point for several compelling reasons. This section delves into why Baofeng is ideally suited for those new to the world of ham radio.

Firstly, affordability is a key factor. Baofeng radios are known for their exceptional value, offering features that outpace their modest price tag. This makes them accessible to a wide audience, ensuring that financial constraints don't hinder anyone's ability to explore amateur radio.

Secondly, ease of use is where Baofeng radios truly shine for beginners. They are designed with the user in mind, featuring straightforward interfaces and essential functions without overwhelming complexity. This simplicity allows newcomers to quickly gain confidence in operating their devices.

The versatility of Baofeng radios is another advantage. Capable of accessing a broad range of frequencies, they provide users with the flexibility to explore different aspects of amateur radio. Whether it's local communication on VHF/UHF bands or experimenting with various ham radio activities, Baofeng radios serve as a gateway to a diverse range of experiences.

Moreover, the Baofeng community plays a significant role in supporting beginners. Online forums, social media groups, and local clubs are filled with Baofeng enthusiasts eager to share their knowledge and experience. This community aspect cannot be overstated, as it provides a supportive environment for learning and growth.

Durability is another hallmark of Baofeng radios. They are built to withstand the rigors of outdoor activities, making them suitable for hikers, campers, and emergency preparedness enthusiasts. This reliability ensures that beginners can count on their Baofeng radios when they need them most.

Programming flexibility is a feature that sets Baofeng apart. With the ability to program channels either manually or via a computer, users have the freedom to customize their radios based on their specific needs and preferences. This adaptability is invaluable as beginners expand their horizons in amateur radio.

Furthermore, the continuous improvement and updates to Baofeng models demonstrate the company's commitment to meeting the evolving needs of amateur radio enthusiasts. This dedication to innovation means that beginners starting with Baofeng can look forward to growing with a brand that stays at the forefront of technology.

Accessories and upgrades are readily available for Baofeng radios, allowing users to enhance their experience as they delve deeper into the hobby. From improved antennas to additional batteries, the options for customization and optimization are plentiful.

For individuals interested in emergency communication, Baofeng radios offer a practical solution. Their dual-band capabilities enable users to monitor emergency frequencies while still being able to communicate on ham radio bands. This dual-purpose functionality is crucial in emergency preparedness and response scenarios.

Additionally, the learning curve associated with Baofeng radios is relatively gentle. Beginners can start with basic operations and gradually explore more advanced features at their own pace. This progressive learning approach prevents the feeling of being overwhelmed, which is crucial for sustaining interest and motivation.

The compact size and portability of Baofeng radios also make them an excellent choice for beginners. Easy to carry on outdoor adventures or store in an emergency kit, these radios embody the principle of readiness and accessibility.

Choosing a Baofeng radio is, in many ways, an investment in personal development within the realm of amateur radio. As beginners progress, the skills and knowledge gained through using Baofeng radios lay a solid foundation for advancing to more complex aspects of the hobby.

Lastly, the sense of achievement that comes from mastering a Baofeng radio cannot be overlooked. For many beginners, successfully making their first contact or participating in a local net is a milestone moment that fuels their passion for amateur radio.

In conclusion, Baofeng radios offer an ideal blend of affordability, ease of use, versatility, and community support, making them a top choice for beginners in amateur radio. As you embark on or continue your journey in this fascinating hobby, Baofeng stands ready to serve as a reliable and empowering companion.

As this section has explored, there are numerous reasons why Baofeng is not just a great option, but perhaps the best first step into the world of amateur radio. Whether your interest is driven by hobby, emergency preparedness, or simply the desire to learn a new skill, Baofeng radios provide the perfect starting point to discover the endless possibilities that await in the realm of radio communication.

Chapter 5:
Radio Fundamentals

Diving into the essence of radio communication opens an intriguing world that's both complex and fascinating, laying the foundation for mastering Baofeng radios and embarking on a fulfilling amateur radio journey. Understanding radio fundamentals is akin to learning the alphabets before stringing sentences; it's essential. This chapter starts by demystifying the various frequency bands, each with its characteristics and applications, making it easier for you to navigate through the airwaves. You'll explore modulation modes including FM, AM, SSB, and more, uncovering how they influence the clarity, range, and overall quality of your communication. Furthermore, we instill in you the basic radio operations and etiquette, ensuring that your transmissions are not just heard but welcomed. Empowering you with this knowledge is not just about enhancing your technical skills; it's about instilling confidence and inspiring you to join the vibrant community of amateur radio enthusiasts. Whether your interest lies in emergency preparedness, outdoor adventures, or simply connecting with people worldwide, mastering these radio fundamentals is a crucial stepping stone. So, let's turn the page and delve into the captivating realm of radio communication, where every discovery and connection contributes to a rewarding and potentially lifesaving hobby.

Frequency Bands Explained

As we dive deeper into the world of radio communication, understanding frequency bands becomes pivotal. Imagine each frequency band as a different lane on a highway, each with its own speed limit and type of vehicle that can travel on it. In radio terms, these lanes are designated portions of the radio spectrum that are allocated for specific uses. Knowing which lane—or band—to drive in, can make all the difference in your journey into amateur radio.

The radio spectrum is divided into many different bands, each characterized by its unique properties. These properties determine how radio waves travel within the band, how they interact with the environment, and what they can be used for. Frequency bands are usually measured in kilohertz (kHz), megahertz (MHz), or gigahertz (GHz), with each band offering distinct advantages and challenges for radio operators.

At the lower end of the spectrum, we find the Low Frequency (LF) and Medium Frequency (MF) bands. These bands, including the familiar AM broadcast band, are characterized by their ability to travel long distances, often around the world, by bouncing off the ionosphere. This makes them ideal for long-range communication, though they are subject to higher levels of noise and interference.

High Frequency (HF) bands, often called the shortwave bands, are where many amateur radio operators spend their time. HF bands offer the ability to communicate over thousands of miles, under the right conditions. This is because HF radio waves can also bounce off the ionosphere and return to Earth, a phenomenon known as "skywave" propagation. These bands are known for their role in international broadcasting, emergency communication, and are heavily populated by amateur radio enthusiasts engaging in global communications.

As we move up in frequency, we encounter the Very High Frequency (VHF) and Ultra High Frequency (UHF) bands. These bands are typically used for local communications over shorter distances. VHF and UHF signals primarily travel by line of sight, meaning they do not typically bend around the curvature of the Earth or reflect off the ionosphere. However, under certain atmospheric conditions, VHF and UHF signals can travel much further than usual; a phenomenon amateur radio operators eagerly exploit.

The VHF band includes the FM broadcast band and segments where you can find marine, aircraft, and land mobile radio services. It's also home to a portion of the amateur radio spectrum used for local contacts and line-of-sight communication. The UHF band, meanwhile, is filled with television broadcasts, mobile phones, satellite communication, and more. For amateur radio operators, UHF offers avenues for exploring digital modes, satellite communication, and terrestrial microwave communication.

Exploring higher up, we find the Microwave bands, stretching from 1 GHz up to 100 GHz. These frequencies are less occupied and provide vast opportunities for amateur radio operators to experiment with point-to-point and satellite communications. Microwave bands are capable of supporting high bandwidth, allowing for high-speed data transmission and cutting-edge communication technologies.

It's essential to note that access to these bands and frequencies is regulated. Amateur radio operators must hold an appropriate license, which grants privileges to transmit on specific bands. These privileges vary according to the license class, with higher license classes granting access to more bands and modes of operation.

Each frequency band has its 'personality,' affected by the time of day, weather, solar activity, and atmospheric conditions. Mastering the use of these bands requires patience, practice, and a bit of experimentation. Amateur radio operators often share their experiences and discoveries, contributing to a collective knowledge base that evolves with the technology and our understanding of radio physics.

Choosing the right frequency band for your communication needs is akin to selecting the right tool for a job. It requires understanding the task at hand and the capabilities of each band. For local communications and emergency preparedness, VHF and UHF bands may be most suitable. For global communication and experimentation, the HF bands offer a vast playground. And for those interested in pushing the boundaries of technology, the microwave bands present limitless possibilities.

In the pursuit of mastering radio communication, embracing the diversity of frequency bands is crucial. Each band offers a unique set of challenges and opportunities to learn and grow. Whether it's making a local emergency call, chatting with a fellow radio enthusiast across the globe, or bouncing signals off the moon, understanding frequency bands is your first step toward unlocking the full potential of amateur radio.

Embark on this journey with curiosity and openness. The world of frequency bands is vast and complex, but also incredibly rewarding. With each band you explore, you'll not only gain practical skills and knowledge but also join a community of passionate individuals who share your interest in connecting and communicating in unique and meaningful ways.

Remember, the journey into the world of radio communication is not just about reaching a destination. It's about the adventure of discovery, the joy of connecting with others, and the satisfaction of mastering a technology that has the power to bring people together, no matter where they are. The frequency bands are your roadmap to this world, offering endless possibilities for exploration, connection, and impact. Embrace them, explore them, and let them inspire you to achieve new heights in your amateur radio journey.

As you continue to explore this fascinating world, keep in mind that each step you take broadens your horizons and deepens your connection to the global community of amateur radio operators. The frequency bands are not just technical resources; they are gateways to adventure, learning, and camaraderie. They are your invitation to be part of a continuing story of innovation, exploration, and communication that transcends boundaries and empowers individuals.

So, let the exploration begin. Let the frequency bands be your guide to a world where communication knows no limits. Where every frequency holds the promise of a new discovery, and every transmission is a step closer to mastering the art and science of radio. Welcome to the journey. Welcome to the heart of amateur radio.

Modulation Modes: FM, AM, SSB, and More

To dive deeper into the world of radio communication, understanding the main modulation modes is essential. These modes—FM (Frequency Modulation), AM (Amplitude Modulation), and SSB (Single Side Band)—each have their unique applications and advantages. As we navigate through these modes, remember, the journey to mastering amateur radio is as exciting as the destinations we can reach with our radios.

Starting with **AM (Amplitude Modulation)**, this is the grandfather of radio modes. Used since the dawn of radio, AM works by varying the strength (amplitude) of the signal to carry information. Though it's not as efficient as newer methods in terms of bandwidth and susceptibility to noise, AM holds a special place for broadcasting music and news on mediumwave and shortwave bands, offering a nostalgic listening experience.

Next, **FM (Frequency Modulation)** steps in with improvements in sound quality and resistance to signal noise. FM varies the frequency of the carrier wave to transmit information. It's the go-to for high-fidelity music stations on the VHF bands and is widely used in two-way radio communication, including most handheld transceivers like Baofeng radios. Its clarity in audio makes it a preferred choice for emergency communication, local radio clubs, and leisure activities.

On a more specialized front, **SSB (Single Side Band)** modulation is a derivative of AM that is more efficient and can travel longer distances. SSB transmits by using only a portion of the AM signal, either the upper or lower sideband, making it the mode of choice for long-distance (DX) communication on the HF bands. It's a bit more complex to tune in to an SSB signal, but the effort pays off with the ability to make contacts across continents.

Exploring further, we encounter **CW (Continuous Wave)**, or more simply, Morse code. This mode, while ancient by today's standards, remains highly effective and efficient. Morse code can cut through noise and reach farther on less power than voice can, making it an invaluable mode for long-distance contacts and emergency situations.

Each modulation mode has a time and place, and choosing the right one depends on your communication needs. For instance, when clarity and quality in local communication are key, FM is your friend. But when you're trying to reach across the globe with limited power, SSB or CW might be your best bet.

Now, in the realm of advanced digital communications, amateur radio isn't left behind. Modes like **FT8**, **PSK31**, and **D-STAR** are taking the stage. These modes use sophisticated digital encoding techniques to send text or voice over long distances, even under challenging conditions. They open up new horizons, making it possible to communicate worldwide with equipment you can hold in your hand or fit on a small desk.

Choosing the right mode is like selecting the right tool for a job. It's not just about the power of your transmitter or the height of your antenna, but also about matching your mode to your mission. Whether it's chatting with a neighbor over FM or reaching out to a distant continent via digital modes, the choice profoundly affects your experience.

As you prepare to embark on your radio adventures, remember that experimenting with different modes is part of the fun. Each offers a unique way to connect and communicate, bringing its own blend of challenges and rewards. It's this experimentation and learning that enrich the hobby, making every contact an opportunity to discover something new.

The beauty of amateur radio lies in its diversity and the vast array of possibilities it offers. Whether you're a hobbyist, a prepper, an outdoor adventurer, or just someone curious about the magic of radio waves, understanding these modulation modes is the first step towards unlocking a world where the airwaves connect us all.

As we close this section, consider modulation modes not just as technical choices, but as gateways to different communities within the amateur radio world. Each mode has its aficionados and its traditions, offering you a chance to learn, to explore, and to belong. From the local FM repeater to the global reaches of SSB and digital modes, there's a place for everyone.

Embrace the journey ahead with curiosity and enthusiasm. Let's not just be operators of radio equipment, but explorers of the airwaves, ever ready to learn, connect, and share. The modes may differ, but the spirit of amateur radio remains the same—a passion for communication that knows no bounds. Armed with knowledge and inspired to experiment, you're now better prepared to dive into the practical aspects of radio operation and to discover the joy of making your first contacts.

Remember, every expert was once a beginner. The world of radio communication is vast and endlessly fascinating. With each step, with each discovery, you're not just operating radios; you're keeping a tradition alive and thriving. A tradition of curiosity, camaraderie, and communication that spans the globe and reaches into the stars.

So, take these lessons as your first steps into a much larger world. A world where, with a simple radio, you can touch the lives of people you may never meet in person, and where your voice can carry far beyond the horizon. Welcome to the world of amateur radio. The airwaves await your call.

Basic Radio Operations and Etiquette

Entering the world of radio communication, especially as a beginner, can seem like navigating a dense forest without a map. However, understanding the basic operations and etiquette of radio communication is akin to finding that map, illuminating the path towards becoming a proficient communicator. This journey is not just about mastering the technical aspects but also about embracing the culture and spirit of the global radio community.

First and foremost, it's essential to comprehend the power behind the push-to-talk (PTT) button. This small action connects you to a vast network of individuals, each with their unique experiences and stories. Yet, with great power comes great responsibility. Before you transmit, always listen. The airwaves are shared resources, bustling with activity. By listening first, you ensure you're not interrupting an ongoing conversation or emergency communication.

Understanding the concept of 'frequency etiquette' is another fundamental aspect. Each frequency band has its unwritten rules about its use. Some are dedicated to emergency services, while others may be popular with radio enthusiasts for casual conversations or 'ragchews.' Researching and adhering to these guidelines helps maintain harmony on the airwaves and ensures you're using the spectrum efficiently and respectfully.

When it comes to initiating communication, the art of calling is critical. A standard procedure involves stating your call sign and sometimes adding 'CQ' if you're looking to make contact with anyone who can hear you. This step is not just about following protocol; it's about making a courteous invitation into the world of radio communication.

Once you've established contact, the essence of radio communication really begins to shine. Exchange information clearly and concisely. It's not just what you say but how you say it. Practicing clear articulation and maintaining an even tone can greatly enhance the quality of your transmissions and make the experience more enjoyable for all parties involved.

However, remember that brevity is the soul of wit. On the air, it's also the key to effective communication. Especially during busy times or when the frequency is in high demand, keeping transmissions short and to the point ensures that others also get their chance to communicate. This practice is not just about etiquette; it's about fostering a culture of mutual respect and cooperation.

Among the most cherished values in the radio community is the spirit of helpfulness and mentorship. Experienced operators, or 'Elmers,' as they're affectionately known, are often eager to lend advice or assistance to those just starting out. Don't hesitate to ask for help or guidance. Embracing this culture of support not only accelerates your learning curve but also enriches the community as a whole.

Equally important is the respect for privacy and discretion. Sometimes, you might accidentally stumble upon personal conversations or emergency communications. In such cases, the ethical path is to refrain from interfering or disclosing any sensitive information you might hear. The integrity of radio operators is a cornerstone upon which the trust and security of the airwaves rest.

Moreover, radio communication isn't just about talking; it's about listening. Active listening can provide you with a wealth of knowledge on how to effectively communicate and operate within various radio environments. It's through attentive listening that you'll pick up on the nuances of different modes and bands, and understand the rhythm and flow of successful radio interactions.

Technical proficiency is undeniably crucial in radio operations. However, emotional intelligence plays an equally significant role. Recognizing and adapting to the tone and mood of the conversation can lead to more meaningful and enjoyable interactions. Whether it's a casual chat or an emergency transmission, empathy and understanding are always key.

Record keeping, while seemingly mundane, is an important aspect of radio operations. Logging your contacts not only helps in maintaining a record for confirmation purposes but also aids in self-improvement and tracking the range and efficacy of your communications. Taking this task seriously can greatly enhance your skills and satisfaction in the radio hobby.

As with any community, conflicts or misunderstandings can arise. Handling such situations with grace and professionalism can help in resolving issues swiftly and maintaining the friendly atmosphere of the airwaves. The goal is always to promote understanding and cooperation, rather than division.

Finally, continuous learning and adaptation are the hallmarks of a proficient radio operator. The world of radio communication is ever-evolving, with new technologies and modes emerging regularly.

Staying informed and open to experimentation can lead to exciting discoveries and personal growth within this fascinating hobby.

In conclusion, mastering basic radio operations and etiquette opens up a world of communication possibilities. It's about much more than just talking into a microphone. It's about joining a global community bound by a passion for connecting across distances, exploring new technologies, and supporting each other. Embrace this journey with enthusiasm, respect, and curiosity, and you'll find that the world of radio communication is an endlessly rewarding adventure.

Remember, every seasoned radio operator was once a beginner. With patience, practice, and perseverance, you too can become a skilled and respected member of the radio community. Let's keep the airwaves alive with the spirit of camaraderie, innovation, and mutual respect. The microphone is in your hands; the world is listening. What will you share?

Chapter 6:
Setting Up Your Baofeng Radio

Now that you've familiarized yourself with the basics of Baofeng radios and understand the fundamental concepts behind radio communication, it's time to propel yourself into the practical world of setting up your own Baofeng transceiver. The journey from unboxing your radio to making your first call is filled with excitement and a bit of technical navigation. Setting up your Baofeng radio is more than a task; it's your initiation into the boundless world of amateur radio communication, where every frequency tuned and button pressed brings you closer to becoming a proficient communicator. This process begins with the initial device setup, inviting you to install the antenna and battery, a simple yet crucial step towards ensuring your device operates at its full potential. Then, you'll embark on a venture through navigating menus and settings, a testament to your growing familiarity with your transceiver's capabilities. Finally, programming your radio, although seemingly a Herculean task, is broken down into digestible basics and tools, ensuring you're not just following instructions but understanding the essence of radio programming. Embrace this chapter as your blueprint towards mastering your Baofeng, igniting a spark of curiosity and confidence as you step into the realm of radio communication, armed with the knowledge and skills to explore, connect, and thrive.

Initial Device Setup and Configuration

Embarking on the journey of setting up your Baofeng radio is a pivotal first step toward unlocking a world rich in communication potential. This process begins with a straightforward yet critical task: assembling the device by installing the battery and antenna, a move that transforms it from a mere object into a gateway of possibilities. Before diving into the vast sea of frequencies and channels, it's essential to familiarize yourself with the radio's basic functions. Navigating through its menus and settings might seem daunting at first, but with patience and curiosity, you'll soon find it to be intuitive. This initial setup is not just about technical steps; it's about laying the foundation for a journey that promises not only to enhance your knowledge but also to connect you with people around the globe. Understanding your Baofeng radio's configuration is your first conquest in this adventure. It's where your theoretical knowledge meets practical skills, offering a taste of the incredible versatility and empowerment that amateur radio provides. Let this process awaken a sense of wonder and determination within you, as each step brings you closer to becoming not just an operator but a true radio enthusiast armed with the power to explore, connect, and conquer challenges. Whether for hobby, emergency preparedness, or exploration, mastering this initial setup paves the way to a world where communication knows no bounds.

Installing the Antenna and Battery

As we venture further into the exciting world of amateur radio, particularly with Baofeng radios, it's crucial to understand that the antenna and battery form the backbone of your communication setup. These components are not just accessories but are essential to your radio's performance and reliability. This section will guide you through the process of installing the antenna and battery on your

Baofeng radio, ensuring you're well-equipped to embark on your amateur radio journey.

First and foremost, let's talk about the antenna. The antenna is your radio's voice to the world. It captures your voice and sends it soaring through the airwaves, reaching fellow enthusiasts, friends, or emergency responders. Installing the antenna on your Baofeng radio is a straightforward process, but it's one that requires attention to detail. Begin by carefully unpacking the antenna from its box. You'll notice that it has a threaded base designed to be screwed onto the antenna connector at the top of the radio.

Gently align the threaded base of the antenna with the antenna connector on the radio. It's important to ensure the threads are properly aligned to avoid cross-threading, which could damage both the antenna and the radio. Once aligned, rotate the antenna clockwise until it's snug. There's no need to overtighten – just a firm hand-tight is sufficient to establish a good connection. Remember, the goal is secure attachment, not proving your strength.

Now, let's shift our focus to the battery – the powerhouse of your Baofeng radio. The battery ensures that your radio remains operational, especially when you're mobile or in situations where charging might not be possible. To install the battery, first, ensure that it's fully charged. This initial charge is vital for battery health and longevity. Match the battery contacts with the contacts on the back of the radio. Slide the battery up towards the top of the radio until it clicks into place, signaling a secure attachment.

With the battery installed, your radio should now power on. However, if it doesn't, don't panic. Double-check to ensure that the battery is properly seated and fully charged. Most issues at this stage are due to simple oversight. Remember, patience and thoroughness are your allies in setting up your Baofeng radio.

Once your antenna and battery are properly installed, it's a good idea to perform a quick functionality test. Power on the radio and scan through the available frequencies. You might not understand everything you hear at this stage, and that's okay. The goal is to confirm that your radio is receiving and capable of transmitting signals. This step is not just about testing; it's about experiencing the immediate result of your setup efforts.

It's essential to recognize that the antenna and battery you've just installed are not permanent fixtures. Over time, you might find the need to upgrade or replace these components. For instance, exploring different antennas can significantly enhance your radio's performance, extending its range and improving signal clarity. Similarly, having spare batteries on hand ensures you're always ready, especially in situations where power sources are scarce.

Understanding how to properly install the antenna and battery on your Baofeng radio empowers you to maintain and modify your setup as needed. Whether you're preparing for an outdoor adventure, participating in emergency communication networks, or simply enjoying the hobby of amateur radio, these foundational skills ensure you're always connected.

Remember, every amateur radio enthusiast was once a beginner. It's through experimenting, learning, and connecting with the community that you'll grow and evolve in this fascinating hobby. As you become more comfortable with your equipment, you'll discover that amateur radio is not just about the technology; it's about the people you meet and the shared experiences along the way.

To inspire you further, consider the antenna and battery as your gateway to the world. With these components correctly installed, you're not just setting up a radio; you're opening a door to endless possibilities. You're stepping into a community that spans the globe, ready to explore, communicate, and connect.

Installing the antenna and battery is the first step in your Baofeng radio journey. Take pride in this accomplishment, for you're not only setting up a piece of equipment, but you're also laying the foundation for a journey filled with learning, discovery, and connection. Embrace it with an open mind and a spirit of adventure.

Remember, as with any journey, there will be challenges along the way. You might encounter technical difficulties, unfamiliar terminology, or even moments of frustration. However, it's through overcoming these challenges that you'll grow. Each obstacle is an opportunity to learn, and with each solution, you become more adept and confident in your abilities.

So, take a moment to reflect on this significant first step. You've successfully installed the antenna and battery on your Baofeng radio, and you're now ready to dive deeper into the world of amateur radio. Whether your interest lies in emergency preparedness, outdoor adventures, or simply the joy of communication, your Baofeng radio is your companion on this journey.

As you move forward, keep exploring, keep learning, and most importantly, keep communicating. The world of amateur radio is vast and varied, offering something for everyone. Your Baofeng radio, equipped with its antenna and battery, is your ticket to this world. Use it to connect, to explore, and to make a difference.

In conclusion, the process of installing the antenna and battery on your Baofeng radio is more than just a technical task; it's the beginning of a larger adventure into amateur radio. With this knowledge, you're well-equipped to not only navigate your Baofeng radio but also to embark on a journey of growth, discovery, and connection. Remember, the world of amateur radio is as vast as it is welcoming. Welcome aboard, and enjoy the journey!

Navigating Menus and Settings

As we journey further into the world of amateur radio, specifically focusing on Baofeng radios, it's pivotal to grasp the art of navigating through its menus and settings. Mastering this aspect will significantly enhance your ability to tweak your device to perfection, aligning it with your unique needs and preferences. Undoubtedly, the plethora of options might seem daunting at first. However, step by step, let's unravel the mystery together, transforming what appears complex into something straightforward and manageable.

The initial step in this process involves turning on your Baofeng radio. Once powered up, you're greeted by a digital display that's your window into the vast capabilities of your device. The 'Menu' button, a portal to customization, is your first point of interest. A single press reveals a list of features, each represented by a numerical code and a descriptor. Don't let this intimidate you. Like learning a new language, familiarity will come with practice.

One of the most fundamental settings to venture into is the frequency mode. Baofeng radios offer the flexibility to switch between frequency mode and channel mode. In frequency mode, you manually input the frequency for the specific channel you wish to access. This mode is invaluable for exploration and when operating on the fly. Dive into this setting with a spirit of experimentation, and remember, toggling back is always an option if you need to return to familiar terrain.

Progressing further, understanding how to adjust the squelch level on your Baofeng can dramatically improve your listening experience. The squelch function serves to mute the background noise when no signal is strong enough to be considered a transmission. Navigating to the squelch setting in the menu and adjusting it helps in clearing out unwanted noise, making sure you catch every word of the crucial communications you're here for.

Another critical setting worthy of your attention is the transmit (TX) power level. As you venture into various environments, from crowded cityscapes to serene landscapes far removed from civilization, the ability to adjust your transmit power becomes essential. Lower power settings conserve battery life and are perfect for short-range communications, while higher settings boost your signal's reach when you're in the vast open or trying to make that all-important connection.

At this juncture, let's not overlook the importance of privacy and security. Your Baofeng radio comes equipped with options to set up privacy tones—CTCSS and DCS codes—that allow your transmissions to be received only by radios set to the same code. This setting, nestled within the menu, is a cornerstone for operating within a team or group, ensuring that your communications remain within your intended circle.

Exploring further, the VOX (Voice Operated eXchange) feature stands out for hands-free operation. This setting, when activated, allows the radio to automatically transmit when you speak, without the need to press the transmit button. It's a game-changer in scenarios where your hands are occupied, ensuring that communication remains seamless and efficient.

As you familiarize yourself with these settings, it's essential to approach the process with patience and a dedication to mastery. Every radio, much like its operator, is unique. The journey to finding the perfect setup is both personal and rewarding. Delve into each menu option, tweaking and testing as you go. It's through this exploration that proficiency is honed, and confidence in your equipment is built.

Moreover, the importance of the dual watch and dual reception features cannot be underestimated. These features enable the monitoring of two different frequencies simultaneously, a critical capability in scenarios where multi-tasking is essential. Engaging with these settings allows for a broader awareness of the radio landscape around you, ensuring that you miss nothing of importance.

Memory channels are another area worth mastering. Programming frequently used frequencies into the memory channels of your Baofeng can save valuable time and streamline your radio use. Whether it's your local repeater, emergency frequencies, or favorite simplex channels, having them stored means accessing vital communications is but a few clicks away.

And let's not forget about the wide range of additional features tucked away within the menu of your Baofeng. Functions like the alarm mode, the FM radio, and the LED flashlight add layers of utility to your device, ensuring it's not just a communication tool but a multifaceted device equipped for a variety of scenarios.

For those looking to dive even deeper, advanced settings such as the R-CTCSS, D-DCS, and offset for repeater access open up the world of amateur radio even further. These settings, while more complex, are the gateway to leveraging your Baofeng to its full potential. They require a deeper understanding but remember, every expert was once a beginner. Approach these settings with curiosity and a willingness to learn.

Lastly, the journey through the menus and settings of your Baofeng is not a solitary one. The amateur radio community is renowned for its spirit of cooperation and mentorship. Online forums, local clubs, and even social media groups are brimming with individuals eager to share their knowledge and experiences. Engaging with these communities can provide invaluable insights, tips, and answers to the inevitable questions that arise.

Remember, every interaction with your Baofeng's menu and settings is an opportunity for learning. Each adjustment, each setting tweaked, brings you closer to becoming not just proficient but proficient and confident in the art of radio communication. Embrace the process, embrace the challenges, and let each step forward fuel your passion for amateur radio.

In conclusion, navigating the menus and settings of your Baofeng radio is a journey of empowerment. It transforms the novice into the adept, the curious into the knowledgeable. It's through understanding the capabilities and intricacies of your radio that you fully unlock its potential, stepping into a world of communication that knows no bounds. So go ahead, power on, press "Menu," and let the adventure begin.

Programming Your Radio: Basics and Tools

Embarking on the journey of programming your Baofeng radio is more than a step towards mastery; it's an initiation into a world where you control your communication destiny. The task might appear daunting at first, but with the right guidance and a bit of patience, you'll soon find yourself navigating the process like a pro. Let's delve into the essentials of programming, equipping you with the knowledge you need to fine-tune your radio to perfection.

First and foremost, understanding the basics of programming is crucial. At its core, programming your Baofeng involves entering specific frequencies, codes, and settings that enable you to communicate effectively across various channels and frequencies. This customization not only enhances your experience but also ensures you're aligned with legal and operational standards.

To embark on this endeavor, you'll need certain tools. The most fundamental of these is the programming cable specifically designed for Baofeng radios. This cable serves as a bridge between your radio and your computer, allowing you to use software to input and adjust settings directly, which can be a more efficient method than manual programming.

Speaking of software, CHIRP is a pillar in the Baofeng programming realm. It's a free, open-source tool that supports a plethora of radio models, Baofeng included. CHIRP's user-friendly interface simplifies the programming process, making it accessible for beginners while also offering the depth required by experienced hobbyists.

Preparation is key before you start. Ensure your radio is fully charged and that you've installed any necessary drivers for the programming cable on your computer. Download and install CHIRP, and familiarize yourself with its layout. Taking these preliminary steps will smooth your path as you move forward.

After connecting your radio to your computer using the programming cable, the next step is to download the current config-uration from your radio into CHIRP. This action creates a safety net, allowing you to revert to the original settings should you need to.

Programming your radio is not merely about adding frequencies. It's about tailoring your device to fit your specific needs and interests in the amateur radio world. Whether it's programming local repeater frequencies, setting up for emergency communication, or experimenting with satellite communications, the possibilities are vast.

Attention to detail is crucial. Each frequency and setting you input must be accurate. Double-check every entry for errors to avoid any operational hiccups. Always reference reliable sources for frequencies and settings specific to your area and interests.

Privacy is another vital consideration. While amateur radio is about open communication, maintaining operational security in certain aspects is advisable. Names, specific locations, and sensitive frequencies should be handled with care and discretion.

After programming, testing your setup is essential. Start with a simple transmission on a frequency you know is active, and seek feedback on your signal's clarity and strength. This iterative approach of tweaking and testing will lead you to optimal settings for your specific needs.

Remember, your radio is a tool for exploration and connection. Programming it effectively opens up new horizons in the vast world of amateur radio. It's about more than just hitting buttons and entering numbers; it's about customizing your journey into the world of radio communication.

As you grow more comfortable with the basics, don't hesitate to explore advanced programming options. From DCS and CTCSS tones to offset settings for repeater access, your proficiency will elevate your radio experience to new heights.

Community resources can be invaluable. Online forums, local clubs, and fellow enthusiasts are wellsprings of knowledge and support. Don't shy away from reaching out for advice or sharing your own insights. The amateur radio community is known for its camaraderie and willingness to help.

Persistence is key. You might not get everything right on your first try, and that's perfectly fine. Each mistake is a learning opportunity, each challenge a chance to grow. Embrace the process with curiosity and resilience, and you'll find that programming your Baofeng radio is a rewarding journey.

In closing, programming your Baofeng radio is an empowering step toward unlocking its full potential. With the basics under your belt and the right tools in hand, you're well on your way to tailor-making your communication experience. Remember, the world of amateur radio is broad and diverse, and how you choose to navigate it is entirely up to you. Let your interests guide you, and let the airwaves be your playground.

Chapter 7:
Antennas and Equipment

Stepping into the world of ham radio brings to light the pivotal role of antennas and the right equipment in enhancing your communication experience. The journey isn't just about owning a Baofeng radio; it's about equipping it with an antenna that promises clarity, range, and efficiency. This chapter breaks down why investing in a quality antenna can be a game-changer, introducing you to the craft of DIY antenna projects that not only save money but also immerse you deeper into the hobbyist's essence of radio communication. Alongside, we'll navigate through the essential accessories that complement your Baofeng, ensuring you're well-prepared for any adventure or emergency situation. Imagine transforming your radio from a mere device into a powerful tool for exploration, connection, and safety. Here, we empower you to make informed decisions on antennas and equipment, guiding you through the technicalities with ease and inspiration. Let's embark on this part of the journey with curiosity and an open mind, ready to embrace the vast possibilities that amateur radio offers.

The Importance of a Good Antenna

In the realm of amateur radio, particularly when starting with versatile and accessible units like Baofeng radios, understanding the pivotal role of a good antenna can't be overstated. While diving into the fascinating world of radio communication, one quickly realizes that the antenna is not just an accessory but the linchpin of effective communication.

Imagine an antenna as the voice of your radio; the clearer and further that voice can travel, the better you can communicate. A well-designed antenna enhances signal strength and clarity, transforming a standard experience into an exceptional one. This is vital for hobbyists who revel in making distant contacts, preppers who depend on reliability in emergency situations, and technology enthusiasts experimenting with the capabilities of their gear.

One might wonder, what makes an antenna good? Several factors come into play—its design, build quality, and suitability for the intended frequency bands and modes of communication. An antenna tailored to specific frequencies can significantly improve both transmission and reception, ensuring your messages are not just sent but also heard.

Moreover, the choice of antenna can greatly influence the range of your Baofeng radio. In scenarios where every kilometer counts, such as emergency preparedness or outdoor adventures, an efficient antenna can be the difference between isolation and connection. It amplifies your ability to reach out over vast distances, often beyond what you might expect with the stock antennas that come with most handheld transceivers.

Yet, the importance of a good antenna extends beyond range. It encompasses the clarity of the transmissions. Interference and static can muddle conversations and hinder crucial communication. A superior antenna cuts through the noise, delivering clear, crisp sound quality that can be a game-changer in both casual conversations among enthusiasts and critical emergency communications.

Another intriguing aspect is the educational journey it offers. Delving into antenna theory and practice encourages a hands-on approach to learning. For those inclined towards DIY projects, constructing and experimenting with different antenna designs can be a rewarding experience, rich with insights into the physics of radio communication.

Experimentation also leads to innovation. The amateur radio community has long been a hotbed for creative solutions, with antennas being no exception. Whether it's tweaking existing models or inventing entirely new configurations, the pursuit of the perfect antenna fuels a spirit of discovery and technical excellence.

This spirit extends to community involvement. Sharing experiences and insights about antenna setups can inspire others to optimize their setups. Forums and local clubs often buzz with discussions on antenna designs, each contributing to a collective pool of knowledge that benefits all members, from novices to seasoned operators.

In emergency scenarios, the antenna's importance is magnified. When traditional communication infrastructures falter, a robust ham radio setup, paired with an efficient antenna, can maintain crucial lines of communication. It's a beacon of hope and a tool of resilience, empowering individuals and communities to withstand and recover from crises.

For outdoor adventurers, an effective antenna means staying connected in the most remote locations. It supports the spirit of exploration, providing a safety net that enables adventurers to push boundaries while maintaining a lifeline to civilization.

Given this array of benefits, investing in a good antenna is investing in your communication potential. This does not necessarily mean spending exorbitantly but rather choosing wisely based on your specific needs, goals, and the unique characteristics of your operating environment.

Crafting or selecting the right antenna is both an art and a science. It involves understanding the underlying principles, the practical aspects of radio operation, and a touch of intuition developed through experience. It's a journey worth embarking on, promising profound satisfaction and continuous learning.

In conclusion, the antenna stands as a central figure in the narrative of amateur radio communication. Its selection, construction, and optimization are not just technical tasks but rites of passage in the journey of a radio enthusiast. A good antenna elevates the experience, bridging distances, clarifying voices, and connecting worlds. It's a testament to the ingenuity and perseverance of those who gaze towards the horizon, seeking to hear and be heard.

As you step further into the world of amateur radio, let the quest for a good antenna guide you. It's a pursuit that encapsulates the essence of the hobby—technical challenge, community, and the sheer joy of making connections. Whether you're reaching out across the globe or simply to a friend a few miles away, remember: it all starts with the antenna.

DIY Antenna Projects

The journey into the world of amateur radio is not just about mastering the airwaves; it's also about embracing the spirit of innovation and creativity. Among the most rewarding ventures you can embark on is crafting your very own antenna. This chapter is dedicated to guiding you through DIY antenna projects that will not only amplify your signal but also your understanding and appreciation of how amateur radio works.

Before we dive into the intricacies of antenna construction, it's important to grasp the essence of why antennas are critical to your radio setup. An antenna acts as the bridge between your radio and the vast network of radio waves traveling through the atmosphere. The right antenna can dramatically improve your radio's performance, extending its reach and clarity of reception and transmission.

Let's start with a project that's perfect for beginners: building a simple dipole antenna. This type of antenna is not only easy to construct but also incredibly versatile. You'll need some basic materials: insulated wire, some connectors suited for your radio model, and a bit of rope for hanging the antenna. The beauty of a dipole lies in its simplicity and effectiveness, making it an excellent first project.

Once you've mastered the dipole, you might want to explore more challenging designs. A Yagi antenna, known for its directional capabilities, could be your next step. Although more complex, constructing a Yagi antenna offers a deeper understanding of radio wave propagation and can be a highly rewarding project. It requires more materials and patience, but the result is a high-gain antenna that can significantly enhance long-distance communication.

For those interested in mobile or outdoor adventures, creating a portable antenna is an exciting challenge. A simple telescopic pole can serve as the basis for a versatile portable antenna that you can take anywhere. Pairing it with a lightweight wire and some basic connectors, you can create a setup that is easy to carry and quick to deploy, perfect for field operations or emergency communication.

In discussing materials, it's not just about what to use but also about understanding why. The choice of wire, for example, can affect the antenna's efficiency and durability. While copper wire is a popular choice for its conductivity and flexibility, you might also consider coated wires for their resistance to weather conditions if your antenna will be exposed to the elements.

Another key aspect of DIY antenna projects is the use of baluns and ununs, devices that help match the impedance of your antenna to your radio. These components can significantly improve the performance of your antenna by reducing SWR (Standing Wave Ratio) and ensuring more efficient transmission and reception of signals.

The process of building your antenna is also an opportunity to familiarize yourself with essential tools and techniques. Soldering, for instance, is a skill that will prove invaluable not just in antenna projects but in many other aspects of amateur radio maintenance and repair.

Testing and tweaking your antenna is an integral part of the DIY process. It's not just about following instructions to the letter; it's about understanding how changes in design and materials affect performance. Using an SWR meter to measure the efficiency of your antenna can guide you in refining your design for optimal performance.

It's also worth considering the integration of your DIY antenna with other equipment, such as amplifiers or tuners. Learning how to optimize this interaction can greatly enhance your radio experience, allowing you to reach frequencies and contacts that were previously out of reach.

Remember, safety first. Working with antennas, especially when installing them at height, carries risks. Ensure you're familiar with safe practices to prevent accidents, and never install antennas near power lines or during adverse weather conditions.

The journey of building your antenna is not a solitary one. The amateur radio community is known for its willingness to share knowledge and experience. Engaging with local clubs or online forums can provide you with valuable advice, inspiration, and maybe even a helping hand.

Embracing the DIY ethos in amateur radio is not just about saving costs; it's a pathway to gaining a deeper connection with your equipment and the science behind radio communication. Each antenna you build is a testament to your growth and curiosity as an amateur radio operator.

In conclusion, DIY antenna projects offer a unique blend of challenge, learning, and satisfaction. Whether you're a novice looking to construct your first dipole or an experienced ham aiming to build a complex directional antenna, the journey is filled with opportunities for innovation and discovery. So, take up the soldering iron, unfurl that wire, and embark on a project that will not only boost your signal but also your spirits.

The world of amateur radio is vast and varied, and with your own hand-built antenna, you're not just listening to the waves; you're becoming an active part of the global ham community. The airwaves await, and your DIY antenna is the key to unlocking their potential. Armed with patience, creativity, and a touch of technical skill, there's no limit to what you can achieve.

Essential Accessories for Your Baofeng

Embarking on your amateur radio journey with a Baofeng handheld transceiver is an exhilarating venture. While the radio itself is a marvel of technology, pairing it with the right accessories can drastically enhance your experience. This chapter is dedicated to guiding you through the essential accessories that will elevate your Baofeng's performance, ensuring you're well-equipped for whatever your radio adventure throws at you.

First and foremost, a high-quality antenna should top your accessory list. The stock antenna that comes with most Baofeng radios performs adequately, but upgrading to a higher-gain antenna can significantly improve both your transmission range and the clarity of received signals. Whether you're communicating across town or trying to hit a distant repeater, a good antenna makes all the difference.

Battery life is another critical consideration for any Baofeng user. Most units come with a standard battery pack, but having an extra high-capacity battery on hand ensures you won't be cut short in the middle of important communications. For those long days out in the field, an extended battery could be your best friend, providing hours of additional operation time.

Considering how your Baofeng will accompany you in various environments, a protective case is a wise investment. A durable case shields your radio from dust, moisture, and the inevitable bumps and drops, extending the life of your device. Some cases offer additional functionality with built-in storage for extra batteries or accessories.

A programming cable is an invaluable tool for any Baofeng owner. While manual programming is possible, a programming cable, coupled with supporting software, simplifies the process of setting up channels and frequencies. This small investment saves you time and frustration, especially when managing multiple channels.

To fully exploit the versatility of your Baofeng, consider acquiring a speaker-mic. This accessory enhances the portability and convenience of your radio, allowing for hands-free operation. Whether you're hiking, biking, or working on a project, a speaker-mic ensures clear communication without needing to hold the radio to your mouth.

For those interested in covert or discrete communications, a good quality earpiece is indispensable. An earpiece not only provides privacy but also clarity in noisy environments. Whether monitoring communications or communicating quietly, an earpiece brings a professional edge to your Baofeng operations.

A reliable external power source, such as a solar charger or a power bank, is crucial for extended outdoor adventures. Being able to recharge your Baofeng's battery when away from conventional power sources ensures you remain connected, no matter where you are.

Experimenting with different antennas can be a fascinating aspect of the ham radio hobby. A magnetic mount antenna for mobile operation or a Nagoya NA-771 for handheld use can open up new horizons. These antennas provide improved performance and the flexibility to explore various aspects of radio communication.

For those interested in maximizing their Baofeng's capabilities, an SMA to SO-239 adapter allows for the connection to external antennas, including high-gain base station antennas. This can dramatically increase the effective range of your handheld radio, especially when operating from a fixed location.

A sturdy belt clip or a hands-free carrying case can make it easier to keep your Baofeng within arm's reach, especially when you're moving around a lot. Mobility and accessibility are paramount for effective communication in various scenarios, from event coordination to emergency situations.

For the amateur radio operator who enjoys tinkering and customizing their setup, a SWR (Standing Wave Ratio) meter designed for VHF/UHF frequencies is a valuable tool. It allows you to test and tune your antenna system for optimal performance, further enhancing your Baofeng's capabilities.

An external microphone can be another excellent addition. This accessory improves audio quality on transmissions, which can be particularly useful in noisy environments or when operating in a vehicle.

Finally, don't overlook the importance of a comprehensive user manual or guidebook. While the Baofeng is user-friendly, delving deeper into its features and functions can unlock its full potential. A detailed manual or a guide focused on Baofeng radios can be an invaluable resource for both beginners and experienced users.

Embarking on the amateur radio journey is not just about owning a radio; it's about creating a versatile, effective communication setup that meets your needs and exceeds your expectations. With these essential accessories for your Baofeng, you're not just prepared; you're empowered to explore the vast, exciting world of ham radio with confidence and clarity. Let your curiosity lead the way, and let these tools amplify your experience, ensuring that every transmission opens a door to new possibilities.

Remember, the goal isn't merely to communicate; it's to connect and thrive in the global ham radio community. Your Baofeng, equipped with the right accessories, is your passport to this vibrant world. Embrace the possibilities, and let your voice be heard across the frequencies. Welcome to the adventure that amateur radio offers, where every day is an opportunity to learn, grow, and connect.

Chapter 8:
On the Air: Making Your First Contacts

Stepping into the world of amateur radio is like opening the door to a vibrant global community, where your first transmission is both a hello and a handshake. Imagine aligning your radio's frequency, your hand poised over the transmit button, ready to bridge distances with the power of radio waves. It's essential to approach your first contacts with a blend of enthusiasm and respect for the established protocols that keep the airways clear and friendly. As you call out, using your newly acquired call sign, the thrill of awaiting a response from an unknown fellow enthusiast somewhere out in the ether is unmatched. This chapter is dedicated to guiding you through the initial steps of making those first connections. From understanding how to properly use your call sign to navigating net operations and adhering to radio etiquette, we've got you covered. Participating in contests and events is not only a great way to test your skills and equipment but also an exciting avenue to engage with the community and make your presence known. Each contact made is a learning opportunity, a chance to refine your operating style, and deepen your understanding of the technicalities of radio communication. Remember, the core of ham radio is communication; every experienced operator once made their first call, and the community is known for its welcoming spirit and willingness to mentor newcomers. So, take a deep breath, key up, and take your first step into the vast and rewarding world of amateur radio. Your journey is just beginning, and the airwaves are alive with potential connections just waiting to be made.

Call Signs and Making Calls

Embarking on the journey of making your first radio contacts can be a venture filled with excitement and a touch of apprehension. It's like unlocking a new level in a quest, where the world opens up through the airwaves, connecting you with fellow enthusiasts from around the corner or across the globe. At the heart of this adventure are your call sign and the art of making calls, both of which serve as your passport and conversational etiquette in the vast community of amateur radio.

Your call sign is more than just a series of letters and numbers; it's your unique identifier, a badge of honor that tells the world you belong to the global fellowship of amateur radio operators. It's crucial to wield it with pride and responsibility, for it carries your reputation across the airwaves. Securing a call sign is a milestone, achieved through obtaining your amateur radio license, and it's a rite of passage that opens the doors to endless communication possibilities.

Making your first call might seem daunting, but remember, every expert was once a beginner. The key is to approach this process with a blend of respect for tradition and openness to learning. Start with listening. Spend time tuning into various frequencies, getting a feel for the rhythm and patterns of exchanges between operators. This will not only familiarize you with the protocols but also build your confidence.

When you're ready to dive in, ensure your equipment is properly set up and tested. A simple, "Is this frequency in use?" can be your initial foray into speaking on the air. It's both a practical check and a polite gesture in the radio community. Upon receiving no answer, you may proceed to introduce yourself using your call sign, followed by a brief message indicating you're seeking contact, such as, "This is [Your Call Sign], a new operator looking for my first contact. Over."

Patience is your ally here. Responses might not be immediate, but the airwaves are often more welcoming than you think. When you do make a connection, the exhilaration is unparalleled. The exchange of call signs follows a rhythm – you share yours, they share theirs, and a bridge of communication forms, enabling a myriad of conversations and learning opportunities.

It's important to note that the art of making calls extends beyond just finding an open frequency and broadcasting your call sign. It's about creating meaningful connections, exchanging information, and often, offering or receiving guidance and mentorship. The frequency bands are vast, each with its own culture and unwritten rules. Navigation through them comes with experience, but every exchange, every conversation, contributes to your growth as an operator.

One of the beauties of amateur radio is the diversity of its community. You might find yourself in a QSO (a conversation) with someone from a completely different background or with a shared interest in emergency preparedness or technology. Each call sign you encounter has a story, a person behind the microphone with experiences and insights to share. This exchange of knowledge enriches not just your technical skills but broadens your horizon on a personal level.

To enhance your calling endeavors, familiarize yourself with the Q-code, a set of standardized codes used by operators. While not mandatory for every conversation, knowing common codes like QSO, QRZ (Who is calling me?), and QTH (location) can streamline communication and demonstrate your commitment to learning the craft.

Moreover, be mindful of etiquette. Respect for others' time and patience in waiting for your turn to speak are golden rules. Amateur radio is as much about listening as it is about talking, and sometimes, the most rewarding experiences come from simply lending an ear to the stories and signals floating through the ether.

While your first few calls might be met with a mix of excitement and mistakes, embrace them as part of the learning curve. Every operator you meet on the air was once in your shoes, and the amateur radio community is known for its welcoming spirit and willingness to guide newcomers.

As you continue to make contacts and build your presence on the airwaves, you'll find that your call sign becomes a reflection of your journey in the world of amateur radio. It will be your introduction in contests, a mark of your contributions during emergency communications, and a signature of friendship in casual conversations.

In conclusion, your call sign is not just an identifier but a key to a global community where every call is an opportunity for growth, discovery, and connection. Embrace the journey of making calls with an open heart and mind. The world of amateur radio is vast and varied, filled with potential lifelong friendships, learning experiences, and moments of triumph. Go forth, express yourself through the airwaves, and become a part of the vibrant tapestry that is amateur radio.

Remember, each call you make is a step towards mastering this fascinating hobby. Every contact is a story, a learning experience, and a chance to make a difference. So, tune in, reach out, and let your call sign echo across the airwaves, marking the beginning of countless radio adventures that await you.

Net Operations and Radio Etiquette

As you embark on your journey into the world of amateur radio, mastering the art of net operations and radio etiquette isn't just recommended—it's essential. A net, or network, in ham radio parlance, is a scheduled gathering of operators on a certain frequency, designed for various purposes, ranging from discussing specific topics to emergency communications. The ability to navigate these gatherings with grace and efficiency speaks volumes about your proficiency as a radio enthusiast.

First and foremost, understanding the structure of a net is critical. Most nets operate with a net control station (NCS), who directs the flow of communication. Think of the NCS as the conductor of an orchestra, ensuring that every participant gets a chance to contribute without the conversation descending into chaos. When you check into a net, always wait for the NCS to call for new stations and provide your call sign clearly when it's your turn.

Listening is just as important as transmitting. By monitoring the net before jumping in, you get a sense of its purpose and tone. Some nets are informal and encourage rag-chewing (ham slang for lengthy, casual conversations), while others are strictly business, focusing on emergency preparedness or directed towards practicing specific communication protocols. Your listening time is valuable; it helps you tailor your contributions appropriately.

Once you're checked into a net, patience becomes your best virtue. Not everyone's equipment is created equal, and some stations might struggle to be heard. It's part of the collective responsibility of the net to ensure that everyone gets a say, regardless of their signal strength. Offering patience and encouragement to fellow operators not only fosters a supportive environment but also solidifies the camaraderie that defines the amateur radio community.

When it's your turn to speak, keeping your transmissions concise and relevant is key. If the net is designed for signal reports, stick to the script, offering clear and useful feedback. For more open-ended discussions, sharing your insights and experiences adds value, but being mindful of time ensures that the net runs smoothly for everyone involved.

Using proper call signs during net operations is not just a matter of etiquette; it's a legal requirement. Always begin and end your transmissions with your call sign, and use the call signs of others when directing comments or questions to specific stations. This practice not only keeps communications orderly but also ensures that everyone on frequency can follow the conversation.

Speaking of order, mastering the phonetic alphabet makes a world of difference in ensuring your call sign is understood, especially under less-than-ideal reception conditions. Articulating "Kilo-Mike-Four-Tango-Bravo-Lima" is far more effective than simply saying "KM4TBL," particularly when static or interference comes into play.

Net operations also offer a unique platform for emergency communication training. Many nets are dedicated to practicing message handling and relay techniques that are crucial during disasters. By participating, you not only sharpen your skills but also contribute to a vital community service, standing ready to aid in times of need.

A fundamental rule of thumb in amateur radio is to never interfere intentionally with other operators. If you stumble upon a net already in progress, take a moment to listen and understand its nature before attempting to join. And, if you accidentally cause interference, a simple and sincere apology is appreciated and respected.

Confidence in your abilities grows with experience, but don't shy away from asking for help or clarifications if needed. The ham radio community is known for its willingness to support and educate newcomers. A question asked today is a mistake avoided tomorrow, and more often than not, fellow operators are happy to provide guidance.

The etiquette extends beyond the microphone. Many nets log their sessions, and following up on these logs online can provide insights into how you're being received by others and offer opportunities for self-improvement. It's also a space to extend gratitude or continue discussions with fellow operators, building relationships that can extend into the real world.

Incorporating these practices into your operating habits will elevate not just your enjoyment of the hobby but also the enjoyment of those with whom you share the airwaves. Remember, every transmission puts your signature on the vast canvas of amateur radio. Making it a positive one ensures that the hobby remains vibrant and welcoming for generations to come.

In conclusion, net operations and radio etiquette form the backbone of the amateur radio experience. They're what transform a sprawling global network of individuals into a cohesive, collaborative community. As you continue to explore and grow within this fascinating hobby, your adherence to these principles will not only mark you as a proficient operator but as a valued member of the amateur radio family.

So, take to the airwaves with confidence and courtesy, armed with the knowledge that you're now well-versed in the nuances of net operations and radio etiquette. Each contact, each net, each transmission is a step on an endless journey of discovery, connection, and, above all, camaraderie. Welcome to the air, and may your signals be strong and your conversations enriching.

Participating in Contests and Events

Jumping into the dynamic world of amateur radio contests and events marks a thrilling chapter in every ham's journey. These activities are not just fun; they're instrumental in sharpening your operating skills, expanding your network, and deepening your understanding of radio mechanics. Whether you're a hobbyist, a prepper, a tech enthusiast, or simply someone enchanted by the magic of radio communication, participating in these events can elevate your experience to new heights.

Ham radio contests, often called 'radiosport', invite operators from around the globe to connect under specific rules and within set timeframes. The objective? It varies from contest to contest but usually involves making as many contacts as possible or reaching hard-to-contact areas. It's a test of skill, endurance, and sometimes creativity, but above all, it's a community event that fosters camaraderie and unity among participants.

For beginners, the thought of jumping into a contest might seem daunting. Fear not. The ham radio community is known for its inclusivity and willingness to mentor newcomers. Start by listening. Tuning into a contest before participating can give you a feel for the pace and etiquette. You'll soon realize that making that first contact is less about competing and more about connecting and learning.

Events aren't limited to contests. They include special occasion broadcasts, commemorative events, and activities designed to advance the art and science of radio communication. Participating in these events can provide you with a richer understanding of the world, history, and the vast potential of radio technology. You'll find events commemorating significant historical moments, celebrating advancements in science, and even promoting global understanding and peace.

One of the first steps in participating is understanding the types of contests and events available. They range from local and regional competitions to global challenges. Each has its own set of rules, objectives, and frequencies. Some focus on specific modes, such as Morse code (CW), while others might concentrate on voice (SSB) or digital modes. The variety ensures that there's something for everyone, regardless of your preferred mode of communication or level of expertise.

Preparing for a contest or event is as much about mindset as it is about technical readiness. Embrace the spirit of learning and exploration. Each contest is an opportunity to push your boundaries, experiment with your equipment, and refine your operating tactics. As you prepare, consider the logistical aspects, such as ensuring your Baofeng radio is programmed correctly, confirming your understanding of the contest rules, and planning your participation schedule.

Technical readiness also involves knowing your equipment inside out. Familiarize yourself with your Baofeng's features, from its frequency range to its power settings. Successful contesting often hinges on optimizing your setup to suit the specific demands of the event. Experimenting with antennas, power sources, and even the physical location from which you operate can significantly impact your performance.

Networking with fellow hams plays a crucial role in enriching your contesting and event participation experience. Joining a local club or online community can provide you with insider tips, firsthand stories, and advice on navigating the complexities of contests. These networks can also offer support, encouragement, and friendship, making your amateur radio journey all the more rewarding.

Maintain a log of your contacts and experiences during contests and events. This isn't just a requirement for contest validation; it's a tool for reflection and growth. Reviewing your log can help you identify areas for improvement, understand patterns in your operating behavior, and measure your progress over time.

The thrill of making that first contact in a contest or event is unmatched. It's a moment of triumph over the airwaves, a confirmation of your skills and preparation. Each contact thereafter adds to your confidence and enjoyment. Whether it's a local competition or a global event, the excitement of reaching out and connecting with someone, somewhere in the world, is a powerful reminder of the magic of ham radio.

Remember, participating in contests and events is not just about winning. It's about learning, growing, and being part of a global community that shares your passion for radio communication. It's about contributing to a legacy of amateur radio that spans decades and touching the lives of people across continents.

As you venture into this exciting aspect of ham radio, approach each contest and event with an open heart and a curious mind. Be ready to learn, willing to make mistakes, and eager to share your experiences. The world of amateur radio contests and events is rich with possibilities, ready to offer you lessons, adventures, and connections that will enrich your journey as a ham operator.

Evolve with every participation. Whether it's tweaking your antenna setup, mastering a new mode of communication, or simply developing patience and strategic thinking, let each contest and event shape you into a more skilled, more connected, and more versatile ham operator.

In the end, the world of ham radio contests and events is not just about signals and frequencies; it's about people. It's a testament to human ingenuity, connection, and the timeless fascination with reaching out across the invisible lines that connect our world. As you embark on this journey, remember that you're not just participating in contests; you're becoming a part of a vibrant, global community of enthusiasts, each with their unique story, connected by the airwaves in pursuit of discovery, communication, and friendship.

So, gear up, tune in, and dive into the exhilarating world of ham radio contests and events. Whether you're seeking to test your limits, expand your skills, or simply make new friends, this journey is sure to offer exciting challenges, rewarding experiences, and infinite opportunities for growth and connection. Welcome to the community, and may your signals reach far and wide.

Chapter 9:
Operating Procedures and Best Practices

Entering the realm of amateur radio operation, especially with Baofeng radios, isn't just about understanding the technicalities; it's about embracing a culture steeped in discipline, respect, and continuous learning. Mastering the art of communication involves a keen awareness of *operating procedures and best practices* that ensure not only successful exchanges but also the safety and privacy of all participants. This chapter dives deep into the essence of smooth, efficient communication protocols, emphasizing the importance of emergency communication protocols that can come in handy during unforeseen circumstances. It also navigates through the often-overlooked aspect of privacy and legal considerations that safeguard you and your fellow enthusiasts from inadvertent regulatory breaches.

While exploring these key areas, remember, each interaction is a chance to learn and grow. The way we communicate, the frequencies we choose, and the respect we show towards the shared spectrum space, speaks volumes about us as operators. Consider this: operating procedures are not just rules; they are the rhythm to which the entire amateur radio community dances. Following them not only enhances your experience but also ensures that the airwaves remain a friendly and welcoming space for everyone. As you gear up to make the most of your Baofeng radio, let these best practices be your guiding stars. They're a testament to the collective wisdom of generations of hams before you, a reminder that in the world of radio communication, we're all learners, and every frequency is a classroom.

Emergency Communication Protocols

In the dynamic world of radio communications, the ability to effectively manage and utilize emergency communication protocols stands as a beacon of responsibility and readiness. When disasters strike or during unforeseen emergencies, the well-honed skills of amateur radio operators can mean the difference between chaos and order, between despair and hope. Embracing this crucial role requires not only technical prowess but a deep commitment to the principles of emergency communications.

Recognizing an emergency situation is the first step in deploying effective communication strategies. Emergencies come in various forms, from natural disasters like earthquakes and floods to man-made crises such as power outages and major accidents. The common thread in these situations is the urgent need for clear, concise, and reliable communication. As operators, your role is to ensure that critical messages are transmitted and received with the utmost efficiency and clarity.

Preparation is the cornerstone of effective emergency communication. This encompasses not only having your equipment in top operational condition but also being well-versed in the protocols that govern emergency communications. Familiarity with national and international guidelines, such as those outlined by the Amateur Radio Emergency Service (ARES) or the Radio Amateur Civil Emergency Service (RACES), is essential. These organizations provide a framework within which amateur radio operators can operate cohesively and effectively in times of need.

A crucial aspect of preparation involves regular practice and drills. Participation in simulated emergency tests (SETs) and other exercises conducted by local amateur radio clubs or emergency management organizations hones your ability to respond swiftly and correctly in real-world scenarios. These drills simulate the pressures and demands of emergency situations, building the resilience and confidence needed to handle actual crises.

The concept of a "go-kit" is a fundamental element of emergency preparedness. A go-kit is a portable, easily accessible pack containing all the essential equipment an amateur radio operator might need in an emergency. This includes not just radios and antennas, but also spare batteries, power sources, a flashlight, and basic first aid supplies among other things. Crafting a comprehensive go-kit ensures that you can remain operational even in the most adverse conditions.

Once an emergency unfolds, effective communication protocols dictate the establishment of a clear command structure. This involves identifying the net control station (NCS) or stations that will coordinate the communication efforts. The NCS acts as a central hub, directing traffic, reducing confusion, and ensuring that critical information reaches its intended recipients. Operational discipline, including proper radio etiquette and adherence to the NCS's instructions, becomes paramount in maintaining an orderly flow of communication.

Understanding and utilizing different frequency bands is an intrinsic part of navigating emergency situations. Different situations may call for different communication bands. For instance, VHF and UHF bands are excellent for local communications, while shortwave (HF) bands can facilitate long-distance communication. Knowing which band to use and how to quickly switch frequencies as needed is a skill that can significantly enhance the effectiveness of emergency communication efforts.

Digital modes of communication, such as packet radio or Winlink, provide valuable tools for transmitting data, including text messages and emails, without the need for a working internet connection. These modes can be especially useful in disasters that disrupt traditional communication infrastructure. Training in these technologies and incorporating them into your emergency communication plan can greatly expand your abilities to relay crucial information.

Interoperability with public service agencies, such as fire, police, and emergency medical services (EMS), enhances the effectiveness of amateur radio communications during emergencies. Establishing relationships with these agencies before an emergency occurs can smooth the path for collaboration during crises. This might involve participating in joint drills, understanding the communication protocols of these agencies, and being ready to provide auxiliary communication support when primary systems fail.

The ethical dimension of emergency communications can't be understated. Privacy, sensitivity to the urgency and content of messages, and the prioritization of critical transmissions over non-essential communication are all ethical considerations that amateur radio operators must navigate. Being a steadfast and reliable communicator also means being a conscientious one, honoring the trust placed in you by both fellow operators and the broader community.

Documentation and debriefing following an emergency operation are critical for continuous improvement. Recording frequencies used, messages relayed, and challenges encountered provides valuable data that can refine future emergency response efforts. Debriefing sessions with all participants can uncover insights and lessons that enhance both individual and collective preparedness for future events.

For amateur radio enthusiasts, technology enthusiasts, and preppers alike, the journey into mastering emergency communication is both a challenge and a profound responsibility. It's a call to harness the power of radio communications to make a tangible difference in times of need, embodying the spirit of service and community that is at the heart of amateur radio culture.

Inspiration can be drawn from the countless stories of amateur radio operators who have provided critical communications under dire circumstances. These stories are not just tales of technical proficiency, but testimonies to the human spirit's resilience, creativity, and unwavering commitment to helping others.

In conclusion, the protocols and practices covered in this section are far more than operational guidelines. They are a reflection of the broader ethos of the amateur radio community—a commitment to preparedness, service, and excellence. As you step into the world of emergency communication, remember that you're not just operating radios; you're part of a global network of individuals dedicated to safeguarding and serving their communities against all odds.

Thus, whether you're a seasoned operator or a newcomer to the world of amateur radio, embracing and mastering emergency comm-unication protocols is a journey worth embarking on. It is an opportunity to blend passion with purpose, technology with humanity, and in doing so, discover the true power of connection.

Privacy and Legal Considerations

Embarking on the journey of amateur radio is not just about mastering the airwaves; it's about understanding the world within which these waves travel. In a landscape where technology and privacy increasingly intersect, amateur radio operators sit at a unique crossroads. This section will delve into the paramount importance of adhering to legal standards and respecting privacy while operating Baofeng radios and other amateur radio equipment.

First and foremost, it's crucial to acknowledge that while the airwaves are vast, they are not lawless. Various regulatory bodies govern the use of radio frequencies to ensure that operators coexist harmoniously and securely. In the United States, the Federal Communications Commission (FCC) sets forth regulations that must be followed diligently. Ignorance of these regulations is not a shield against accountability. Hence, staying informed and compliant is the mark of a responsible operator.

Privacy on the airwaves is an intriguing concept. Unlike digital communication, which often comes shrouded in layers of encryption, radio transmissions are inherently open to anyone with the right equipment to intercept. This openness requires a sense of responsibility among operators to exercise caution and respect. Sharing sensitive personal information or that of others can not only breach privacy but also lead to unintended consequences.

Intersecting with privacy is the issue of encryption. The allure of encrypting transmissions to ensure privacy can be strong. However, it's essential to understand that, according to FCC regulations, transmitting encrypted messages on amateur radio bands is prohibited. The spirit of amateur radio is communication and community, not secrecy. The limitations on encryption aim to maintain the transparency and openness that define the amateur radio ethos.

Recording conversations is another area where privacy and legality intertwine. Before pressing the record button, an operator must be aware of consent laws which can vary significantly by location. In some jurisdictions, all parties involved must consent to the recording, making it crucial to communicate intentions clearly and secure agreement from all involved.

The use of call signs in amateur radio serves several purposes: it identifies the operator, contributes to the organization of the airwaves, and, importantly, helps in maintaining accountability. Using your call sign correctly is not just a matter of legal compliance; it's a gesture of respect towards the amateur radio community and the norms that sustain it.

It's also vital to navigate the nuances of interference. While unintentional interference can occur, deliberate interference, or "jamming," is both disruptive and illegal. Understanding how to minimize your impact on others' transmissions and how to diplomatically resolve interference issues you may encounter is a core skill for any operator.

The aspect of modifications to radios, including Baofeng models, is a topic ripe with enthusiasm. Yet, even in the pursuit of enhanced performance, legal considerations must govern modifications. Unauthorized changes that result in a device operating outside its licensed parameters can have serious legal repercussions and potentially harm the broader radio ecosystem.

Outside the technical realm, there's a social dimension to legal and privacy considerations. Foster a culture of respect and integrity within the amateur radio community. Encourage dialogue about privacy, share knowledge about legal matters, and lead by example. The strength of the community lies in its collective respect for these principles.

Regarding emergency communication, the lines between necessity and privacy can blur. In disaster situations, the imperative to communicate critical information swiftly must be balanced with the discretion that privacy demands. Mastering this balance is a skill that underscores the role of amateur radio operators in emergency preparedness.

Moreover, when discussing sensitive topics or transmitting information that could be classified as private, employing generic terminology and avoiding specifics can mitigate privacy concerns.

Developing an awareness of what should and should not be shared on the airwaves is essential.

Engaging in international communications introduces another layer of complexity. Different countries have their own rules and norms about radio operation and privacy. When your signals cross borders, so too do the legal and ethical considerations. Familiarizing yourself with international agreements and respecting the privacy norms of different cultures is a mark of a worldly and wise operator.

Educational content is a treasure trove for any amateur radio enthusiast. However, sharing or disseminating content without proper authorization can infringe on copyright laws. Respecting intellectual property rights extends to the choice of material broadcasted over the airwaves.

Ultimately, navigating the intertwining paths of privacy and legal considerations demands continuous learning and adaptability. The landscape is ever-changing, with technological advancements and shifts in legislation. Staying informed, engaged with the community, and committed to ethical operation are your best strategies for a rewarding journey in amateur radio.

In conclusion, the realm of privacy and legal considerations in amateur radio is vast and complex, yet deeply fascinating. It challenges operators to think critically about their responsibilities and the impact of their actions. By embracing these challenges with a sense of curiosity and duty, you contribute to a culture of respect, legality, and community in the amateur radio world. Let your journey be guided by these principles, and you'll find that the airwaves are not just a space of freedom but of profound connection and respect.

Chapter 10:
Advanced Operating Techniques

Embarking on the journey from understanding the basics to mastering advanced operating techniques in ham radio is both an exhilarating challenge and a profound opportunity for personal and community growth. As we delve into this next phase, you'll discover the power of using repeaters to magnify your voice across vast distances, a technique that transforms obstacles into gateways for global communication. The exploration of satellite communications opens up a celestial frontier, where the sky is not a limit but a canvas for innovation and connection. Additionally, the realm of digital modes and Slow Scan Television (SSTV) offers a unique blend of traditional radio with the cutting-edge digital world, enabling you to send images and data across the airwaves. This chapter aims not just to impart knowledge, but to inspire you to push the boundaries of what is possible with your Baofeng radio. As you hone these advanced skills, remember that each step forward is a leap towards becoming a more adept communicator, a valuable member of the amateur radio community, and a pioneer in leveraging technology for enriching connections and safeguarding our collective future.

Using Repeaters to Extend Range

Embarking on the journey into advanced operating techniques, we breach the topic of utilizing repeaters to magnify the reach of your communications. Repeaters are not merely tools; they are beacons that bridge vast expanses, serving as the backbone for extended radio communications among amateur radio enthusiasts. Whether for hobbyists, preppers, or outdoor adventurers, understanding and utilizing repeaters can drastically transform your radio experience.

At their core, repeaters are essentially signal boosters strategically placed on high ground — often on towers, tall buildings, or mountain peaks — to receive transmissions on one frequency and simultaneously retransmit them on another. This retransmission allows signals to cover distances far beyond their original capability, overcoming natural obstacles and the curvature of the Earth.

For newcomers wielding a Baofeng radio, the concept of connecting to a repeater might appear daunting at first glance. Yet, with a blend of instruction and empowerment, mastering this skill will unveil a new horizon of possibilities in your radio communication endeavors. It's about taking that initial step of curiosity to exploratory action.

Setting the scene, imagine you're in a lush forested valley, surrounded by an embrace of greenery and the challenge of rugged terrain. Direct communication from one end to the other may be hindered by the very beauty that surrounds you. Here, repeaters come into play, serving as the invisible bridge that connects your message through the obstacles imposed by nature.

The process begins with programming your Baofeng radio. While each model may have its nuances, the principle remains consistent: you'll need to input both the repeater's input and output frequencies along with any required access tones - specifically, the Continuous Tone-Coded Squelch System (CTCSS) tone or Digital Code Squelch (DCS) code. These tones are the secret handshake between your radio and the repeater, ensuring your signals are warmly received and broadcasted onward.

Finding a repeater to use might initially seem like seeking a lighthouse in a stormy sea. However, resources abound. Local clubs, online databases, and dedicated apps provide comprehensive listings of repeaters, complete with frequencies, tones, and coverage maps. It's about tapping into the existing framework of the amateur radio community, a treasure trove of shared knowledge and collaboration.

The beauty of amateur radio is its spirit of perpetual learning and adaptation. Experimenting with different repeaters, adjusting settings, and fine-tuning your approach are parts of the iterative process that hones your skills. Don't shy away from reaching out within the community for advice or to share your discoveries. Growth in this realm is communal, with every individual's advancement contributing to the collective wisdom.

It's crucial, however, to approach repeater usage with respect and mindfulness. Repeaters are communal resources, often maintained by clubs or individual enthusiasts at their own expense. Observing proper etiquette—identifying yourself, pausing to allow for emergency traffic, and expressing gratitude—reflects well on you and fosters a positive environment for everyone.

Engaging with repeaters is more than a technical endeavor; it's a venture into a world where distances shrink, and communities grow closer. Whether it's making new contacts, participating in nets, or simply listening and learning from the exchanges that flow through these airwaves, repeaters offer a gateway to enriching experiences.

What's exhilarating is the scalability of knowledge and skills in the use of repeaters. As you grow more comfortable and proficient, you can delve into the nuances of setting up your own repeater, contributing an essential asset to the amateur radio community and experiencing the profound satisfaction of facilitating communication for others.

Yet, the use of repeaters isn't just about extending range for the sake of connection. In emergency preparedness and response scenarios, the ability to communicate effectively over extended distances can be a lifeline. It underscores the significance of amateur radio not just as a hobby, but as a critical component in disaster readiness and community safety.

In stepping back, it's apparent that employing repeaters in your amateur radio practice isn't just an advanced technique; it's a path to becoming a more integrated, active participant in the global amateur radio community. It's about contributing, learning, and, above all, connecting in ways that defy the limitations of geography and terrain.

As you venture forward, let the use of repeaters be a beacon guiding you towards greater heights in your radio journey. It's a testament to human ingenuity and the spirit of connection that defines the heart of amateur radio. Embrace the challenge, revel in the learning, and expand the scope of your communications beyond what you thought possible. Let each transmission be a step towards not just greater reach, but deeper connections and shared adventures in the world of amateur radio.

With each frequency dialed, each tone set, you're not merely operating a radio; you're weaving yourself into a vast network of voices, stories, and lives interconnected through the airwaves. So, take heart, take action, and let the world hear your call. The journey is yours to undertake, and with repeaters in your toolkit, the horizon is boundless.

Introduction to Satellite Communications

Embarking on our journey through satellite communications opens up a new dimension of amateur radio that might initially seem intimidating but is thrilling and immensely rewarding. Satellite communications enable hobbyists to reach beyond the horizon, connecting with people and places in a way that is unfathomable with traditional radio signals. It's a bridge between the fundamental concepts we've grasped and the vast expanse of the cosmos, waiting for us to explore. In this section, we'll delve into the essentials of satellite communications, unfolding the mystery and showcasing how you can leverage this advanced operating technique to enhance your amateur radio experiences.

The essence of satellite communications lies in using artificial satellites as repeaters in the sky. Unlike terrestrial repeaters perched on high buildings or mountaintops, satellites orbit the Earth, receiving signals transmitted from one location and rebroadcasting them to another. This remarkable ability allows communication over enormous distances, crossing continents and oceans with ease. It challenges us to think bigger, to extend our reach beyond the visible horizon and touch the stars.

Understanding the orbits of satellites is crucial. Their pathways around the Earth determine their visibility from a specific location and influence the planning of communication sessions. Some satellites, known as low earth orbit (LEO) satellites, zoom around the planet quickly, offering brief windows for communication. These require precise timing and planning to utilize effectively. However, the thrill of making that fleeting connection offers a rush of excitement that is hard to match.

Another critical aspect of satellite communications is the Doppler effect. As satellites move relative to the Earth, the frequency of the signals they transmit appears to shift. This phenomenon requires us to adjust the frequencies we use to maintain communication. It introduces an intriguing challenge, blending scientific principles with the art of radio communication and adding another layer of engagement to our hobby.

For amateur radio operators, satellite communications open up a realm filled with opportunities for exploration and learning. It's not just about the technical aspects. It's about being part of a global community that looks upwards, striving for connections that were once the domain of professional astronauts. With a simple handheld transceiver and a home-built antenna, you can tap into this world, experiencing the exhilaration of satellite QSOs (contacts).

Starting with satellites might seem daunting, but remember, every expert was once a beginner. It's about taking that first step with curiosity and enthusiasm. The investment in learning and equipment pays off immensely, not just in the contacts you make but in the personal satisfaction of overcoming challenges and continuously learning.

The tools required for satellite communications are surprisingly accessible. While high-end gear can enhance the experience, many amateurs start with nothing more than a dual-band handheld radio and a simple directional antenna. This minimalist setup is enough to dive into the world of satellite communications, proving that the barriers to entry are not as high as one might think.

Selecting the right satellite to communicate with is your next step. Various amateur satellites, each with unique characteristics, are orbiting the Earth right now. These include FM repeater satellites, linear transponder satellites, and digital satellites. Each type offers a different kind of experience and requires different operating techniques. Beginning with FM satellites is a common and relatively straightforward entry point into satellite communications.

Antenna tracking is another fascinating part of satellite communications. As satellites move across the sky, adjusting your antenna to maintain a strong connection becomes a dynamic and engaging activity. It connects us physically to the process, turning dial tweaking and frequency adjustments into a dance of coordination and timing.

Moreover, satellite communications are not just a personal pursuit. They are a shared journey, connecting us with a community of like-minded individuals. Amateur radio clubs and online forums abound with enthusiasts who are eager to share their knowledge and experiences. These communities offer support, advice, and sometimes, life-long friendships. They remind us that amateur radio, in all its forms, is ultimately about connection.

For the adventurous spirit, satellite communications offer a gateway to something greater. They remind us of the incredible technological achievements humanity has made and our potential for innovation and exploration. They're a testament to our curiosity and desire to reach beyond our grasp, to communicate over vast distances and to touch the sky. Whether you're a seasoned ham operator or just starting, the world of satellite communications welcomes you with open arms and infinite possibilities.

As we conclude this introduction to satellite communications, remember that every journey begins with a single step. This isn't just about mastering a technical skill; it's about embracing adventure, patience, and continuous learning. The challenges you'll face are stepping stones, leading you to moments of triumph and realization. So, take that step, join the global conversation, and let the sky be your playground.

With practice and perseverance, your initial foray into satellite communications will evolve into a passion. The sense of achievement when you make your first satellite contact is unparalleled. It's an affirmation of your skills, a testament to your ability to navigate complexities, and a celebration of your place within the vast, interconnected world of amateur radio.

In the following pages, we dive deeper into the nuts and bolts of satellite communications, breaking down the concepts, technologies, and operating procedures that will guide you from curiosity to competency. Remember, the journey might seem long, but every QSO, every challenge overcome, and every new piece of knowledge acquired is a step toward mastering this majestic facet of amateur radio. Welcome to the wondrous world of satellite communications, where the sky is not the limit—it's just the beginning.

Exploring Digital Modes and SSTV

In the adventure of building your proficiency in amateur radio, venturing into digital modes and Slow Scan Television (SSTV) introduces a fascinating realm of possibilities. This transition from the traditional analog communication forms brings us closer to the frontier where innovation meets tradition. Let's embark on a journey to demystify these concepts, fostering a deeper connection with fellow enthusiasts worldwide.

Digital modes in amateur radio represent a shift towards using digital signals to communicate. This isn't just an evolution; it's a revolution in how messages are transmitted and received. These modes, including FT8, JT65, PSK31, and RTTY, enable you to communicate across vast distances, even under challenging conditions where voice or Morse code might falter. The beauty lies in their efficiency and the minimal bandwidth they occupy, making every transmitted bit of information count.

Let's not overlook Slow Scan Television (SSTV), which is like a cherry on top of the digital modes' sundae. SSTV allows the transmission of images over the radio waves, turning what was once a solely auditory experience into a multimedia adventure. Imagine receiving a picture from halfway across the globe, all transmitted through the airwaves. It's like receiving a postcard through frequency rather than through the post.

To dive into digital modes and SSTV, it's paramount to grasp the basic technical requirements. You'll need a computer or a digital interface connected to your radio. Software applications designed for these modes encode and decode the digital signals. The magic happens here: what you type or the images you select are converted into tones and transmitted over the airwaves, to be decoded back into text or images by the receiving station.

Starting with digital modes can feel like learning a new language. And indeed, it is. Each mode has its unique characteristics, advantages, and sometimes, its specific purpose or preferred use scenario. For instance, JT65, with its ability to pull signals out of the noise, is perfect for weak signal communication. On the other hand, PSK31, favoring real-time chats, might be your go-to for more 'conversational' digital communication.

The allure of SSTV lies in its simplicity and the profound impact of visual communication. Setting up SSTV is similar to other digital modes but uses different software specifically designed for sending and receiving images. The process might seem elaborate initially, but the reward of seeing an image materialize on your screen, piece by piece, is incredibly satisfying.

Experimentation is key in mastering digital modes and SSTV. Each interaction, each transmission, and each received signal is a learning experience. Start with simpler modes like PSK31 or FT8 and gradually work your way up to more complex or less common modes. The online amateur radio community is an excellent resource for advice, software recommendations, and inspiration.

Operating in digital modes and SSTV also underscores the importance of etiquette. Remember, the frequencies used for these modes are shared resources among the global amateur radio community. Adhering to best practices ensures that everyone can enjoy these modes without stepping on each other's transmissions.

For those drawn towards adventure, exploring digital modes and SSTV is akin to setting sail into uncharted waters. Every QSO (contact) teaches something new, every received image adds a layer to your experience, and every challenge overcome is a testament to your growing skills.

The intersection of amateur radio with modern digital technology exemplifies how the hobby evolves with time, integrating new technologies to enhance the experience. Your journey through these modes isn't just about the contacts you'll make or the images you'll receive; it's about being part of a tradition of continuous learning and innovation.

Moreover, digital modes and SSTV hold a unique position in emergency communication. They can transmit crucial information efficiently when traditional communication systems fail. In an emergency, being able to send detailed instructions, maps, or images can be a game-changer for response teams.

To those embarking on this journey, remember: patience and persistence are your allies. There will be hiccups and learning curves, but the amateur radio community is known for its supportive nature. Don't hesitate to ask for advice, share your experiences, and help others along the way.

In conclusion, the exploration of digital modes and SSTV opens new horizons for amateur radio enthusiasts. It's an invitation to blend traditional skills with new technologies, to connect in novel ways, and to keep the spirit of innovation alive within the community. So power up your station, tune into the digital frequencies, and let the pixels and packets flow. The digital realm awaits, full of potential contacts and undiscovered aspects of amateur radio to explore.

As we move forward in this book, let's carry the spirit of exploration and curiosity with us. The journey through amateur radio is as much about the connections we make with others as it is about the personal growth and knowledge we gain. Whether you're decoding your first digital signal or receiving an image from across the globe, remember: every beep, every pixel, every transmission is a step towards becoming a more skilled and connected amateur radio operator.

Chapter 11:
Maintenance and Troubleshooting

In the journey of mastering radio communication, learning the art of maintenance and troubleshooting your Baofeng radio is as essential as the initial thrill of making your first contact. Think of your radio as a trusted companion, one that requires care to perform at its best. This chapter will guide you through basic maintenance practices that not only prolong the life of your device but also enhance its performance. Understanding how to keep your device in top condition, from regular cleaning to battery care, can prevent common issues that might otherwise interrupt your communication endeavors. Yet, even with meticulous care, you might encounter hitches. Here, we delve into the most common problems you might face, such as poor reception, battery issues, or software glitches, and offer practical solutions to get you back on the airwaves efficiently. We'll also touch upon firmware upgrades and adjusting advanced settings, empowering you to customize your device for a tailored communication experience. By equipping yourself with these troubleshooting skills, you become more resilient and adept, ready to tackle challenges as you continue to explore the vast and rewarding world of amateur radio.

Basic Maintenance for Longevity

Maintaining your Baofeng radio isn't just about ensuring it works; it's about valuing the connection it represents. A well-maintained radio is your ticket to a world of vibrant communication, a lifeline in emergencies, and a steadfast companion on your amateur radio journey. In a world accelerating towards the future, taking care of your devices is a nod to the enduring value of human connection.

First and foremost, cleanliness is key. It might seem trivial, but keeping your radio clean from dirt, dust, and moisture can drastically extend its life. Use a soft, dry cloth to gently wipe its surface. For tougher spots, a slightly damp cloth can do wonders, but ensure no moisture makes its way inside the device. This simple routine can prevent numerous technical issues.

Battery care is equally critical. Lithium-ion batteries, like those in your Baofeng, prefer a partial rather than a full discharge. Avoid running your radio's battery down completely before charging; instead, charge it when it reaches about 40% capacity. This method can significantly enhance the battery's lifespan, ensuring that it remains a reliable source of power for your communications.

Antenna maintenance also deserves attention. Ensure the antenna is securely fastened and remains upright during operation. If you notice any bends or damage to the antenna, it's wise to replace it promptly. A functional antenna is crucial for clear communication and will help in transmitting your signal further without unnecessary strain on your radio.

Next, consider the storage of your device. When not in use, store your Baofeng radio in a cool, dry place away from direct sunlight. Extreme temperatures and sunlight can degrade the battery and the radio's external materials. Proper storage ensures your radio remains in optimal condition, ready for your next adventure or emergency situation.

Software updates are often overlooked but essential for maintenance. Stay current with the latest firmware updates for your Baofeng radio. These updates can fix bugs, improve performance, and sometimes unlock new features. Regularly check the manufacturer's website or online forums dedicated to Baofeng radios for information on the latest updates.

Let's talk about the speaker and microphone ports. These are vulnerable points where dust and moisture can easily get inside your device. Keep them clean and dry. A soft brush can remove any accumulated dust, and a dry, cool blast of air can help dislodge any particles stuck inside.

Do not overlook the importance of the programming cable and software. They are crucial tools for your radio's maintenance, allowing firmware updates and frequency programming. Keep the cable and software in good condition by handling them gently and keeping them updated, respectively. An issue with these tools can lead to communication problems down the line.

Regularly practicing with your radio is an often-ignored form of maintenance. Familiarity with its functions means you're more likely to notice when something's amiss. This intimate knowledge of your equipment empowers you to preemptively address issues before they escalate into more significant problems.

Accessories like the belt clip and protective case can also play significant roles in maintenance. They protect your radio from physical damage, which can lead to internal issues. Invest in quality access-ories to keep your device in good shape, ensuring it can withstand the rigors of your adventures.

Handling your radio with care cannot be overstated. Simple actions like avoiding dropping it, not exposing it to water or harsh chemicals, and not over-tightening the antenna can prevent most hardware issues. Respect your device as a sophisticated piece of technology that's more than just a tool—it's your gateway to the global ham radio community.

For those who venture into the great outdoors with their radio, be mindful of environmental conditions. Humidity, salt water, and dirt can be particularly harsh on electronic devices. In such environments, extra precautions like using waterproof bags or cases are advisable to protect your radio.

The connection and accessory ports on your Baofeng radio require periodic inspection as well. Ensure that these points are clean and free from debris. A stable connection is essential for charging and programming your radio, and neglect here can lead to frustrating connectivity issues.

Documentation of your radio's performance and any maintenance actions you've taken can be incredibly valuable. Keep a log to track battery life, signal strength, and any anomalies. This record can help you pinpoint issues more rapidly and might even assist others who encounter similar problems.

In summary, maintaining your Baofeng radio is about more than just technical care—it's about honoring your gateway to global communication. Through simple, regular maintenance, you're not just extending the life of your device; you're ensuring that it remains a reliable, powerful tool in connecting with the world. Let your Baofeng radio be your steadfast companion in your amateur radio endeavors, adventure gatherings, and emergency preparedness. With proper care, it will serve you faithfully, amplifying your voice across the globe.

Common Issues and Solutions

Embarking on the journey of amateur radio, especially with a Baofeng radio, is an exhilarating experience. It's a blend of tradition and technology that can open up untold avenues for communication and exploration. However, as with any technological endeavor, you might encounter bumps along the road. Fear not, for every common issue encountered, there's a solution at hand.

One of the most frequent challenges new operators face is poor reception. This often stems from using the stock antenna that comes with the Baofeng radios. While functional, these antennas are not always the best for clarity and range. The solution? Investing in a higher quality antenna. This small change can significantly improve your reception quality and expand your communication range, transforming your radio experience.

Another regular hiccup involves programming difficulties. Navigating the Baofeng's menu system can be daunting for beginners. Many find manual programming to be a challenging task. The remedy to this is twofold: patience and practice. Familiarize yourself with the manual, step by step. Additionally, consider using programming software. This tool drastically simplifies setting up frequencies and channels, easing your programming woes.

Battery life can also be a concern. Some users report that their Baofeng radios deplete batteries quicker than expected. To counteract this, ensure you're using a battery at its optimum health. Moreover, carrying spare batteries or investing in a larger capacity battery extends your operational time, ensuring you're never cut off mid-conversation.

Squelch settings are often overlooked, yet they play a crucial role in your listening experience. If your squelch is set too high, you might miss weaker signals. Conversely, if it's too low, you'll hear a lot of background noise. Finding that sweet spot where you receive clear transmissions without unnecessary noise can take some tinkering, but once found, it enhances your listening quality immensely.

Sometimes, you might encounter an issue where your radio turns on, but there's no sound. Before assuming the worst, check to see if the VOX (Voice Operated Exchange) feature is accidentally activated. If it is, the radio may be in a transmit mode by mistake, muting all reception. Deactivating VOX can promptly resolve this issue.

When facing transmission difficulties, inspecting your antenna connection is paramount. A loose or poorly connected antenna can severely impact your ability to send out strong, clear signals. Ensure the antenna is screwed in securely and that there's no damage to the antenna or the connector on the radio.

Audio quality issues might not always be due to hardware problems. Sometimes, incorrect settings, like the Narrowband/Wideband setting, can affect your sound. Make sure this setting matches the standard used by those you're communicating with. A simple adjustment can clear up audio quality significantly.

Frequency errors are a common pitfall. Double-checking that you're on the right frequency and that there's no inadvertent offset applied is a good practice. Misunderstandings in frequency settings can lead to unsuccessful attempts to connect with repeaters or other operators.

For those venturing into the realm of digital modes or software-defined radio (SDR) links with your Baofeng, compatibility issues can arise. Ensuring your software is up to date and correctly configured for use with your specific Baofeng model can mitigate many of these problems.

Interference is an inevitable part of the radio experience, yet there are strategies to minimize its impact. Identifying the source of interference and adjusting your frequency or location can often alleviate the issue. Additionally, using filters or adjusting your squelch settings can help block out unwanted noise, improving your overall experience.

Understanding privacy settings and frequencies is vital. Operating within legal frequencies and respecting privacy guidelines ensures you're not inadvertently eavesdropping or broadcasting where you shouldn't be. Familiarizing yourself with FCC rules and regulations is not just about compliance; it's about protecting the integrity of the radio community.

Software glitches occasionally occur, rendering your radio unresponsive or erratic. Performing a factory reset can often resolve this. Remember, resetting your radio will erase all your settings, so this should be a last resort. Always back up your configurations if possible.

The journey of learning and solving these common issues is intrinsic to the radio hobby. It builds not just troubleshooting skills but resilience and adaptiveness. Embrace each challenge not as a setback, but as an opportunity to deepen your understanding and proficiency.

Lastly, remember that the amateur radio community is remarkably supportive. Online forums, local clubs, and fellow enthusiasts are invaluable resources. They can offer advice, insights, and sometimes, the simplest solutions that might not have crossed your mind. Engaging with the community can transform troubleshooting from a solitary task into a collective journey of discovery.

By addressing these common issues with patience, curiosity, and a bit of elbow grease, you're not just maintaining your Baofeng radio; you're enhancing your skills as a radio operator. Each challenge conquered is a step forward in your amateur radio adventure, paving the way for clear skies and clear frequencies.

Firmware Upgrades and Advanced Settings

Embarking on the journey of amateur radio is akin to setting sail into a vast sea of discovery, communication, and technology. Among the critical aspects of this adventure is understanding the importance of firmware upgrades and advanced settings, especially for Baofeng radio enthusiasts. It's not just about keeping your radio up-to-date; it's about unlocking a new realm of possibilities and optimizing your device for peak performance.

Firmware, in essence, is the software programmed into the radio, controlling its functions and features. Think of it as the soul of your Baofeng radio. Over time, manufacturers release firmware updates to fix bugs, add new features, or enhance existing ones. Upgrading your firmware ensures your radio remains a reliable tool, whether for everyday communication or in emergency scenarios.

Initiating a firmware upgrade may seem daunting, especially for those new to amateur radio. The first step is to identify your radio's current firmware version. This can typically be done by accessing a specific menu on your device, often found in the settings. Knowing your version is crucial, as it helps determine the necessity and compatibility of an upgrade.

Once you've established that an upgrade is needed or beneficial, the next step is to visit the manufacturer's website. Here, you'll find the most recent firmware versions along with instructions on how to download and install the update. It's paramount to follow these instructions meticulously to avoid any mishaps that could potentially incapacitate your radio.

Understanding the advanced settings of your Baofeng radio can equally enhance your amateur radio experience. These settings allow you to fine-tune your radio to your specific needs and the unique demands of your operating environment. From adjusting squelch levels to setting up privacy codes, these advanced operations enable you to communicate more effectively and secure your transmissions.

One particular setting that often goes underutilized is the Dual Watch feature. This capability allows users to monitor two frequencies simultaneously, vastly broadening your communication horizon. Imagine the convenience of listening to an emergency frequency while still being connected to a local repeater, ensuring you never miss critical information.

Another advanced feature worth exploring is the menu access to VOX (Voice Operated Exchange) settings. VOX enables hands-free operation by detecting when you start speaking and automatically transmitting your voice. Perfect for situations where manual push-to-talk operation is inconvenient or impossible.

Customizing the power settings on your Baofeng can also have a significant impact on your operating experience. Many users don't realize that they can conserve battery life by reducing the transmission power when communicating over short distances. Conversely, for those hard-to-reach contacts, increasing the power can give your signal the necessary boost to get through.

Beyond the practical enhancements, venturing into the world of firmware upgrades and advanced settings can be incredibly rewarding. It's a testament to your growth as an amateur radio operator and your commitment to mastering your craft. Each new feature learned and applied is a step towards becoming a more proficient, versatile communicator.

Tackling firmware upgrades and delving into advanced settings also fosters a deeper connection with the global amateur radio community. Sharing experiences, solutions, and discoveries with fellow enthusiasts not only enriches your own knowledge but contributes to the collective wisdom of the community.

It's worth mentioning that while pursuing these upgrades and settings, caution and patience are virtues. Rushing through firmware updates or hastily adjusting settings without understanding their impact can lead to frustration and setbacks. Always proceed with deliberate attention to detail and, when in doubt, seek advice from more experienced operators.

The realm of firmware upgrades and advanced settings is where technical skills and creative problem-solving converge. It's where you learn to think like an engineer, adapting and customizing your device to fit your unique amateur radio profile. Embracing this aspect of radio maintenance and troubleshooting not only enhances your capabilities but also adds a rich layer of enjoyment to the hobby.

As you navigate through the ever-evolving landscape of amateur radio, remember that mastery comes with time and experience. Firmware upgrades and exploring advanced settings are an integral part of that journey. They represent the continuous innovation and adaptability at the heart of amateur radio. It's a pursuit that challenges you, rewards you, and always keeps you learning.

In the grand adventure of amateur radio, every new setting adjusted, every firmware upgrade, is a step closer to realizing the full potential of your Baofeng radio. It's about more than just communicating; it's about thriving in a world interconnected by the airwaves. It's a pursuit that not only prepares you for the challenges of tomorrow but also connects you to a community of like-minded explorers today.

So, take that step. Dive into the world of firmware upgrades and advanced settings. Your journey into the depths of amateur radio is only just beginning, and every discovery, every adjustment, is a treasure waiting to enhance your experience. Embrace the challenge, revel in the learning, and watch as you transform not just your radio, but your connection to the world around you.

Chapter 12:
Joining the Ham Radio Community

Embarking on the journey of becoming a seasoned ham radio operator isn't just about mastering the technical aspects; it's equally about weaving yourself into the fabric of the ham radio community. This chapter aims to guide you through the exhilarating process of connecting with fellow enthusiasts, finding your tribe within clubs and organizations dedicated to amateur radio, navigating the vast ocean of online resources and forums, and immersing yourself in the spirit of hamfests and conventions. Ham radio clubs are not just about antennas and transceivers; they are the heartbeat of the amateur radio world, offering mentorship, friendship, and a shared passion for communication. Online forums and resources serve as your 24/7 portal to advice, troubleshooting help, and the latest in amateur radio trends. Meanwhile, attending hamfests and conventions is akin to attending the greatest show on earth for radio enthusiasts, offering unparalleled opportunities to learn, shop, and bond with like-minded individuals. As you dive into this chapter, envision yourself not just as an individual with a radio, but as a vital part of a global community that thrives on connection, innovation, and the sheer joy of reaching out and touching the world through the airwaves.

Clubs and Organizations for Enthusiasts

Embarking on your amateur radio journey introduces you to a world brimming with opportunities, not just in the airwaves but through the vibrant community of ham radio enthusiasts. This chapter focuses on the very essence of amateur radio's spirit – the clubs and organizations that knit the global community together. These entities are not just groups; they're the lifeblood of the ham culture, offering support, knowledge, and comradery.

The moment you step into the ham radio world, you'll find an inclusive community that's eager to welcome newcomers. Clubs are the cornerstone of this community, providing a structured platform for learning, mentorship, and exchange of ideas. They vary widely in their focus and activities, from local clubs that meet regularly in person to virtual groups that connect enthusiasts from across the globe.

Joining a local amateur radio club is your doorway to merging theoretical knowledge with practical skills. These clubs frequently host workshops, licensing classes, and "Elmer" sessions—the ham radio term for mentoring. Through these, you'll learn not just the technical ins and outs of your Baofeng radio but the etiquette and joy of making your first contact.

But clubs aren't just about getting on the air. Many are deeply involved in community service, providing emergency communications during natural disasters or public events. Participating in these activities not just sharpens your operating skills but embeds a sense of purpose in your hobby. It's a profound way to give back to the community using your newfound skills.

On an international scale, organizations such as the American Radio Relay League (ARRL) in the United States serve a pivotal role in the amateur radio realm. The ARRL not only provides a wealth of resources for both beginners and seasoned hams but also advocates for amateur radio rights at the legislative level. Being part of such an organization connects you with a global network of enthusiasts and keeps you informed on the latest in amateur radio.

Moreover, special interest groups within the ham radio community cater to nearly every interest. Whether you're into digital modes, satellite communication, or QRP (low power) operation, there's a group for you. These groups are fantastic for deepening your knowledge in specific areas and connecting with individuals who share your passion.

Engaging with these clubs and organizations also opens the door to participating in contests and awards. These activities aren't just about showcasing your skills but are a fun and exciting way to enhance your abilities, understand propagation, and even design your own antennas.

The aspect of mentorship that clubs and organizations offer can't be understated. Ham radio is a hobby where experience is prized, and the community loves to uplift its members. The guidance from seasoned hams can help you navigate not just technical challenges but also understand the culture and etiquette that's central to amateur radio.

For those with a competitive streak, clubs often participate in "Field Day" events, which are equal parts emergency preparedness exercise and social gathering. Field Days are an excellent way to test your skills under unconventional conditions, often operating off-grid and constructing makeshift antennas. It's learning, competition, and fun rolled into one event.

Don't underestimate the social aspect of these organizations. The friendships and connections you make through clubs and organizations often last a lifetime. Amateur radio is one of those rare hobbies where your network is both global and local, and friendships are based on shared enthusiasm and experiences.

Navigating through the vast options of clubs and organizations might seem daunting at first, but it's worth the effort. Start with local clubs that are easily accessible and branch out as your interests develop.

Remember, the goal is to find a community that supports your journey, offers resources, and aligns with your interests.

Online forums and social media groups also play a significant role in today's amateur radio community. They offer platforms for asking questions, sharing experiences, and connecting with other enthusiasts. While they can't replace the hands-on learning and fellowship of in-person clubs, they're invaluable resources for information and support.

In conclusion, clubs and organizations are at the heart of the amateur radio experience. They provide education, resources, community service opportunities, and a sense of belonging. As you delve deeper into the world of ham radio, let these groups guide you, inspire you, and connect you with the fascinating and diverse world of amateur radio enthusiasts.

Remember, your journey in ham radio is your own, but you're never alone. The community is always here, ready to welcome you with open arms, share in your excitement, and support you through challenges. Embrace the opportunities that clubs and organizations offer, and let them enrich your amateur radio adventure.

So, take that step, reach out, and join a club or organization. It's more than just a membership; it's a gateway to a world of knowledge, adventure, and friendship. Let your amateur radio journey truly begin.

Online Resources and Forums

As you embark on your journey into the world of ham radio, immersing yourself in the vast ocean of online resources and forums can dramatically accelerate your learning curve and enhance your experience. Imagine having an entire community at your fingertips, ready to provide insights, answer questions, and offer encouragement as you navigate through the complexities of radio communication. The Internet is a treasure trove of information, and savvy enthusiasts know how to tap into this wealth to broaden their knowledge and skills.

Online forums dedicated to ham radio are particularly invaluable for beginners. These platforms offer a space where you can ask even the most basic questions without fear of judgment. Whether you're struggling with programming your Baofeng radio, looking for advice on the best antenna to use, or simply want to understand the nuances of radio etiquette, there's always someone willing to help. More experienced hams often frequent these forums, sharing their wisdom and learning from the collective experiences of the community.

Aside from forums, there are numerous websites and blogs that cater to amateur radio enthusiasts. These online resources can serve as comprehensive guides, covering everything from the fundamentals of radio communication to advanced operating techniques. They often provide step-by-step tutorials, product reviews, and DIY project ideas that can inspire and challenge you to expand your capabilities.

YouTube channels and online courses offer another dimension of learning, providing visual and auditory learners with detailed demonstrations and explanations. Seeing someone perform a task or explain a concept in a video can sometimes make a world of difference in understanding. These resources are particularly helpful for hands-on activities such as setting up a radio station, programming a radio, or building antennas.

Social media groups and pages are also fantastic for connecting with other ham radio enthusiasts. These platforms offer a more casual setting for sharing experiences, asking questions, and staying updated on the latest news and events in the amateur radio world. Many of these groups also organize virtual meetups, offering a chance to connect with hams from around the globe.

Podcasts focused on amateur radio can be a great way to absorb information while on the go. Listening to episodes featuring interviews with experts, discussions on the latest trends, and advice on overcoming common challenges can provide valuable insights and keep you entertained during your daily commute or while performing other tasks.

Email newsletters from reputable amateur radio organizations can provide a condensed digest of the most important news, articles, and opportunities for learning and involvement. Subscribing to these newsletters means you won't miss out on important updates, contests, and events that could enrich your ham radio experience.

For those looking to dive deep into the technical aspects, online databases and archives offer access to a wealth of research papers, manuals, and schematics. This level of detail is indispensable for understanding the science behind radio communication and for troubleshooting or modifying equipment.

While the abundance of information available online is certainly beneficial, it's important to approach online forums and resources with a critical eye. Not all advice is good advice, and it's wise to cross-reference information to ensure it's accurate. However, don't let this discourage you. Questioning and verifying information is part of the learning process and is essential for growth.

Engaging with online communities also offers the opportunity to give back. As you gain experience and knowledge, contributing to forums and social media groups not only helps others but reinforces your own understanding. Teaching is, after all, one of the best ways to learn.

The camaraderie found in the ham radio community is something truly special. The willingness to help, the shared excitement for new discoveries, and the mutual respect for knowledge and experience create an environment that fosters growth and learning. Online resources and forums are at the heart of this community, providing a platform for connection and education that transcends geographical boundaries.

In conclusion, as you journey through the world of ham radio, let online resources and forums be your compass and guide. They offer a path to knowledge, a platform for connection, and a means to deepen your involvement in this fascinating hobby. Whether you're seeking answers, inspiration, or camaraderie, the online amateur radio community welcomes you with open arms. Embrace the opportunities that these resources provide, and you'll find yourself not just as a participant in the world of ham radio, but as an active and contributing member of its vibrant community.

Remember, the journey of a thousand miles begins with a single step. Let your curiosity and passion drive you, and don't be afraid to seek help and ask questions. The world of ham radio is expansive and ever-changing, and there's always something new to learn. With the wealth of online resources and forums at your disposal, you're well-equipped to navigate this exciting journey. So go ahead, dive in, and start exploring. The ham radio community awaits.

Hamfests and Conventions

Joining the ham radio community introduces you to a world where technology meets tradition, and where the spirit of exploration and innovation is ever-present. One of the most exciting aspects of being part of this community involves participating in hamfests and conventions. These events serve as vibrant meeting grounds for amateurs, enthusiasts, and professionals alike, offering an invaluable opportunity to learn, engage, and grow in the field of amateur radio.

Hamfests, at their core, are gatherings where ham radio enthusiasts come together to exchange equipment, ideas, and experiences. They're like fairs, but with a focus on radio equipment and technology. Here, you might find yourself rummaging through tables of vintage radios, the latest transceivers, antennas, and various electronic parts. It's akin to treasure hunting for the radio aficionado, where the thrill of the hunt is as rewarding as the find itself.

Conventions, on the other hand, tend to be more formal and structured. They often feature presentations from leading figures in the field, workshops on various aspects of amateur radio, and discussions on the latest trends and technologies. They provide a fantastic opportunity for learning and professional development within the hobby. These events are also where regulations, advancements, and the future of amateur radio are debated and shaped.

For beginners, attending a hamfest or convention can be an eye-opening experience. It's here that the theoretical aspects of amateur radio come to life. You're not just reading about antennas or frequencies; you're seeing them, touching them, and talking to people who've mastered their use. These events can significantly accelerate your learning curve and deepen your appreciation for the hobby.

The camaraderie and sense of community at these gatherings are unparalleled. Ham radio is a hobby that transcends age, backgrounds, and borders, uniting people through a shared passion for communication and technology. Engaging in conversations, sharing stories, and exchanging knowledge with fellow enthusiasts can inspire and motivate you in ways you might not anticipate.

Another highlight of hamfests and conventions is the opportunity to participate in live demos and workshops. These hands-on experiences allow you to see how different equipment operates in real-world scenarios and learn about setup, maintenance, and trouble-shooting directly from experienced operators.

It's also worth noting that many of these events host licensing exams. If you've been preparing for your license but haven't taken the step to take the exam, a local hamfest might be the perfect occasion to do so in a supportive and encouraging environment.

One of the best parts about hamfests and conventions is the access to equipment. Whether you're in the market for new gear, looking for rare components to complete a project, or just browsing, the variety and deals you can find at these events are often unmatched. It's also a great place to sell or trade your equipment and find buyers who appreciate the value of what you're offering.

To get the most out of hamfests and conventions, it's a good idea to go in with a plan. Review the schedule, highlight workshops or talks you don't want to miss, and set aside time for browsing the vendor areas. Don't be shy to engage with exhibitors and presenters - the amateur radio community is known for its welcoming and helpful nature.

Don't overlook the smaller, local hamfests either. While they may not have the scale of national conventions, they offer a more intimate setting for networking and learning. These events can be particularly valuable for newcomers looking to establish connections within the local amateur radio community.

For those who've delved deeper into the hobby, hamfests and conventions offer opportunities to specialize further. Whether it's satellite communications, digital modes, or emergency preparedness, you'll find like-minded individuals and specialized sessions catering to your interests.

Networking at these events can also open doors to new opportunities. Many ham radio enthusiasts have found mentors, joined clubs, and formed lasting friendships through connections made at hamfests and conventions. It's a reminder that even in a hobby as technical as amateur radio, the human element plays a pivotal role.

In preparation for attending your first hamfest or convention, try to get your hands on a programmable Baofeng radio. These affordable, versatile devices are perfect for experimenting with different aspects of amateur radio. Familiarizing yourself with a Baofeng beforehand can make attending workshops and demos at the event even more rewarding.

Finally, remember that hamfests and conventions are not just about buying, selling, or learning – they're celebrations of the amateur radio spirit. They embody the curiosity, ingenuity, and passion that drive the hobby forward. Whether you're a newcomer eager to make your first radio contact or a seasoned operator looking to explore new frontiers, these gatherings remind us of the endless possibilities that amateur radio offers.

Embrace these opportunities with an open mind and a willingness to engage. The experiences, knowledge, and friendships you'll gain at hamfests and conventions can profoundly enrich your journey in the ham radio community.

Chapter 13:
Ham Radio and Emergency Preparedness

As we've delved into the world of Baofeng radios and the broader universe of amateur radio, it's become clear that ham radio isn't just a hobby; it's a lifeline in times of need. This chapter, "Ham Radio and Emergency Preparedness," is more than a guide—it's a call to action for every amateur radio enthusiast to recognize their potential role as key players in emergency communication. By building a comprehensive emergency communication kit, you're not just preparing yourself; you're shouldering a responsibility towards your community. Understanding the pivotal role of amateur radio during disaster scenarios can transform the way we perceive readiness. It's not merely about having the right equipment; it's about harnessing knowledge and skills to make a difference when seconds count. Additionally, volunteering with emergency response organizations doesn't just put your skills to good use; it embeds you in a network of like-minded individuals committed to safeguarding their communities. Each aspect of preparation, from selecting the right gear to participating in drills, is a step towards becoming more than a hobbyist. You're becoming part of a global network of first responders, ready to provide critical communication links when traditional systems fail. This chapter is your blueprint for evolving from an amateur radio enthusiast into a prepared, capable communicator who can stand in the gap during crises, embodying the spirit of service and resilience that defines the ham radio community.

Building an Emergency Communication Kit

When it comes to being prepared for unforeseen disasters or emergencies, having an emergency communication kit isn't just an option—it's a necessity. Within the realm of ham radio and Baofeng radios, this kit takes on an even more critical role, enabling you not just to keep in touch with the world during times of crisis, but also to provide assistance to those in need. This section aims to guide you through assembling a comprehensive emergency communication kit, ensuring you're ready for anything that comes your way.

First and foremost, the cornerstone of your kit should be a reliable ham radio. Baofeng radios are a popular choice for beginners due to their affordability and ease of use. Select a model that suits your needs, keeping in mind the importance of durability and battery life in emergency situations. Ensure you have a deep understanding of how to operate your chosen device, as this knowledge becomes invaluable during a crisis.

Alongside your radio, a supply of batteries is essential. Consider investing in rechargeable batteries and a solar-powered charger. This setup not only reduces your reliance on the power grid but also ensures that you can maintain communication capabilities over an extended period, even when traditional charging methods aren't available.

An extra antenna, particularly one designed for increased range, can be a game-changer in an emergency. High-gain antennas can significantly extend your radio's reach, potentially connecting you with individuals far outside your immediate area. This capability might be crucial for receiving assistance or coordinating with emergency services.

Waterproofing your equipment is another critical step. Emergencies often come hand in hand with adverse weather conditions. Ensuring your radio and accessories are protected from the elements can mean the difference between staying connected and being completely cut off. Waterproof bags and cases are widely available and can offer peace of mind during tumultuous times.

A comprehensive instruction manual, or better yet, a quick reference guide for your radio, should be an integral part of your kit. Even if you're well-versed in operating your radio under normal circumstances, the stress of an emergency can make it difficult to recall specific functions or settings. Having a guide to refer to ensures you can quickly and efficiently utilize your radio when it matters most.

It's also wise to include a list of essential frequencies and local repeaters in your kit. This list can help you find help and information more quickly during a disaster. Familiarize yourself with these frequencies beforehand, but keep the list on hand for easy access when under stress.

Headphones or an earpiece for your radio can make communication clearer, especially in noisy environments. This simple addition can significantly improve your ability to both send and receive critical information during a disaster.

Don't forget the importance of personal safety and comfort. Your communication kit should include basic survival items such as a first-aid kit, flashlight, multitool, and emergency blanket. While not directly related to communication, these items can help you sustain until help arrives or until the situation stabilizes.

Maintaining your kit is as important as building it. Regularly check the condition of all items, replace batteries as needed, and update your frequency list to reflect any changes in local repeater information. This ongoing maintenance ensures your kit is always ready to go at a moment's notice.

Lastly, practice makes perfect. Regular drills using your emergency communication kit can uncover any gaps in your preparation or knowledge. These practice runs can help familiarize yourself with your equipment under pressure, making you more effective in real-world scenarios.

Building an emergency communication kit is an empowering step towards self-sufficiency and resilience in the face of adversity. It embodies a proactive approach to preparedness, ensuring you're equipped not just to survive, but also to assist others during times of crisis. The process of assembling and maintaining your kit also deepens your understanding of ham radio operations, making you a more proficient and confident operator.

In conclusion, the creation of an emergency communication kit is a journey that enhances your readiness for any situation. It allows you to harness the power of ham radio, transforming you from a mere bystander to an active participant in emergency response. Embrace this journey with dedication and enthusiasm, for the skills and knowledge you acquire along the way could one day make a significant difference.

Remember, in the world of emergency preparedness, communication is key. You now have the blueprint to build an emergency communication kit that brings together technology, know-ledge, and a proactive mindset. With this kit, you're not just preparing for the unknown; you're stepping into a role of responsibility and empowerment, ready to face whatever challenges may come with confidence and clarity.

Role of Amateur Radio in Disaster Scenarios

When traditional communication infrastructures crumble under the overwhelming force of natural or man-made disasters, the value of amateur radio shines brightest. Known affectionately as ham radio, this medium of communication has proven time and again to be a lifeline in scenarios where others fail. In an era dominated by digital communication, the resilience and versatility of amateur radio in disaster scenarios stand as a testament to its importance.

Ham operators, equipped with their radios, can establish networks that cut through the silence of a blackout. When cell towers fall and internet services falter, these passionate individuals bridge the gap, providing critical communication links for emergency services, disaster relief organizations, and even affected communities themselves. The immediacy with which these networks can be established makes amateur radio an essential tool in any emergency preparedness kit.

Moreover, the global community of amateur radio enthusiasts operates on a foundation of goodwill and a commitment to serve. This spirit of service becomes most visible during crises. Hams around the world have often been among the first to relay information out of disaster zones, and they have a storied history of providing real-time data and logistical support that saves lives.

The versatility of amateur radio also extends to its ability to operate on multiple frequencies and modes. This flexibility means that operators can adapt to the specific needs of a disaster scenario, whether that involves coordinating rescues, passing messages from isolated locations, or supporting existing emergency communication efforts.

Training and drills play a critical role in preparing amateur operators for disaster scenarios. Many local clubs and organizations regularly participate in simulated emergency tests and real-world disaster drills. These exercises, often conducted in collaboration with government and non-government organizations, ensure that amateur operators are ready to deploy their skills when the unexpected occurs.

The importance of amateur radio in emergency communication is further underscored by its inclusion in national emergency preparedness strategies. Many governments recognize the invaluable role that amateur operators play and provide support in the form of licensing reliefs during emergencies, making it easier for hams to get on the air and serve their communities.

Case studies of amateur radio's role in disaster response abound, from earthquake relief efforts in Haiti to hurricane recovery in the Caribbean and wildfire containment efforts in Australia and California. In each scenario, amateur operators provided critical communication channels, often being the first to report on conditions and needs from the ground.

Another aspect that highlights the role of amateur radio during disasters is its ability to provide emotional support to affected individuals. Being able to communicate with loved ones or just hearing a friendly voice over the radio can be a huge morale booster for those caught in devastating circumstances.

Furthermore, amateur radio's role in disaster scenarios is not just limited to the immediate aftermath. In the weeks and months that follow, as communities rebuild, ham radios often facilitate long-term recovery efforts, including logistics coordination and community support services.

Technology advancements have also augmented the capabilities of amateur radio. Incorporation of digital modes and internet-linked communication systems allows for wider reach and more robust data transmission, making it an even more powerful tool during disasters.

Yet, despite these advancements, the core of amateur radio's value in emergencies remains its human element. The skills, resourcefulness, and dedication of amateur operators make all the difference. It is their ability to innovate, adapt, and persevere that truly defines the role of amateur radio in disaster scenarios.

To those new to the world of amateur radio, the prospect of contributing meaningfully to disaster response efforts can serve as a powerful motivator. It's an opportunity to combine a fascinating hobby with the potential to make a significant impact on the lives of others in times of need.

As we look to the future, the role of amateur radio in disaster scenarios is likely to grow even more important. With climate change and urban expansion increasing the frequency and severity of disasters, the demand for flexible, reliable communication solutions will only rise. Amateur radio, with its unique blend of technology, community, and resilience, is exceptionally well positioned to meet this challenge.

In conclusion, the role of amateur radio in disaster scenarios is invaluable. As both a tool and a community, it embodies the best of human ingenuity and compassion. For those drawn to amateur radio, either out of interest in the technology or a desire to contribute to emergency preparedness, the potential to make a real difference is immense. It's a testament to the power of communication and the enduring spirit of service that defines the amateur radio community.

Volunteering with Emergency Response Organizations

In the realm of amateur radio, the spirit of service combines seamlessly with the love of technology and communication. Volunteering with emergency response organizations is not just about providing a set of skilled hands during disasters; it's about becoming a part of a larger, global effort to save lives, and preserve safety and security. It's here that amateur radio operators step into the light, blending their passion with purpose.

Being an amateur radio operator offers a unique opportunity to serve communities in times of need. Emergency response organizations, including ARES (Amateur Radio Emergency Service), RACES (Radio Amateur Civil Emergency Service), and other local disaster response teams, constantly seek skilled volunteers who can operate ham radios. This skill set becomes crucial when traditional communication networks are compromised or completely down.

When you volunteer with these organizations, you're not just signing up to transmit messages. You're becoming part of a trusted team that may be activated during natural disasters like hurricanes, earthquakes, wildfires, or during man-made emergencies like large-scale power outages. It's about having the ability to provide a critical communication link, relaying vital information between the public and emergency services.

Getting involved starts with training. Many emergency response organizations offer workshops and exercises to prepare volunteers for real-world scenarios. This training covers not just ham radio operations but also the protocols and etiquette for emergency communication, ensuring that when the time comes, your transmissions are clear, concise, and above all, effective.

The beauty of volunteering lies in the flexibility it offers. You can participate at a level that matches your availability and comfort. Whether it's playing a role in emergency exercises, offering your expertise in logistics and network setup, or being on the frontline during disasters, every bit of help strengthens the community's resilience.

It's also about continuous learning and improvement. After each deployment, teams review their performance, looking for areas of improvement. These debriefings are invaluable, as they not only enhance the effectiveness of the team but also boost your skills and understanding of emergency communications.

Moreover, technology in emergency communication is always evolving, and as a volunteer, you will have the opportunity to work with cutting-edge equipment and techniques. This might include digital modes of communication, software-defined radios, or satellite communications, all of which can provide reliable alternatives when traditional methods fail.

One of the most inspiring aspects of volunteering is the community you become part of. The camaraderie and bonds formed within these groups are strong, forged in shared experiences and a common goal of serving others. It's a community that supports its members, encourages growth, and welcomes anyone willing to learn and contribute.

For those interested in making a difference, the first step is reaching out to local emergency response organizations or amateur radio clubs. Many clubs have a dedicated liaison for emergency communications and can guide you on how to get started, what training is required, and how you can fit into their existing frameworks.

Remember, the role of an amateur radio operator in emergencies goes beyond technical skills. It's about being calm under pressure, making quick decisions, and adapting to rapidly changing situations. These are qualities that are honed over time, through training, practice, and real-world experience.

Volunteering with emergency response organizations also offers a unique perspective on the importance of preparedness, both at the individual and community level. It's a reminder of the vulnerability of our modern communication infrastructure and the need for robust, alternative means of staying connected during disasters.

In conclusion, volunteering is a fulfilling way to apply your interest and skills in amateur radio toward a noble cause. It's about being ready to step up when the unexpected happens, ensuring that through your actions, you're helping to keep people safe, informed, and connected. It's a responsibility that comes with challenges but also with immense rewards.

As you consider adding this dimension to your amateur radio journey, remember that it's about more than just radios and frequencies. It's about people, communities, and the impact you can have. It's a path that offers growth, learning, and the satisfaction of knowing that you're part of a global network of volunteers, ready to make a difference when it matters most.

Conclusion

As we close the pages of this journey into the world of amateur radio and Baofeng radios, it's essential to reflect on the path we've traversed together. From deciphering the basics of ham radio, delving into the historical evolution that shaped its existence, to the intricate details of getting licensed and mastering the operation of Baofeng radios, this guide has sought to arm you with not just knowledge but the confidence to dive into radio communication.

The allure of amateur radio is timeless, bridging generations with waves that traverse the expanse of our skies. It's a hobby that brings together diverse individuals, a tool for emergency preparedness, and an avenue for endless learning and discovery. Your journey into ham radio, powered by the versatility and accessibility of Baofeng radios, marks the beginning of an adventure that promises to be as enriching as it is thrilling.

Getting licensed, although initially seeming daunting, is your first step towards unlocking the vast potential that amateur radio holds. Remember, the license is more than just a permit; it's your ticket to a global community of enthusiasts, ready to welcome you with open frequencies. The importance of a good antenna, the thrill of making your first contact, and the satisfaction of participating in contests and events, all build the foundation of an engaging and fulfilling hobby.

The chapters dedicated to setting up your Baofeng radio, understanding antennas and equipment, and advanced operating techniques were crafted to demystify the technical aspects of radio operation. These sections are not mere instructions; they are your toolbox for innovation, experimentation, and exploration in the radio communication domain.

Emergency preparedness and the role of amateur radio in disaster scenarios underscore the profound impact ham radio can have on society. Your ability to communicate when traditional means fail could mean the difference in critical situations, highlighting the significance of this hobby far beyond recreation.

Joining the ham radio community, participating in clubs, and engaging with online forums offer more than just shared knowledge and resources; they provide a sense of belonging, a family of like-minded individuals bound by their passion for radio communication.

Maintenance, troubleshooting, and continuous learning underline the dynamic nature of amateur radio. The landscape of radio communication is ever-evolving, with new technologies, regulations, and challenges emerging. Staying informed and adaptable is key to not just sustaining but thriving in this hobby.

Your journey, equipped with a Baofeng radio, might start with curiosity but unfolds into a lifetime of learning, discovery, and connection. Every frequency tuned, every contact made, and every challenge overcome enriches your experience and contribution to the ham radio community.

The chapters preceding this conclusion have laid out a foundation, a launching pad into the vast expanse of radio communication. But remember, the real adventure lies beyond these pages, in the hands-on application, experimentation, and exploration you'll embark on.

Let this book be a reminder that the journey into amateur radio is not just about mastering technical skills but about embracing a hobby that fosters global connections, lifelong learning, and a resilient community. Whether for hobby, emergency preparedness, or exploration, the world of amateur radio and Baofeng radios opens up a universe of possibilities waiting to be discovered.

As you turn off the final page, don't see it as an end but as a beckoning horizon inviting you to step into the world of amateur radio. Let your curiosity lead the way, your passion fuel your journey, and your Baofeng radio be your guide. The frequencies are your canvas; paint them with your unique callsign, making your mark in the vast network of amateur radio enthusiasts.

Embrace the challenges, celebrate the successes, and cherish the connections you'll make along the way. Amateur radio is not just a hobby; it's a journey, an adventure, and a community. And now, with this book as your guide, you're ready to take your first steps into a world where every frequency holds a story, every contact a new friend, and every challenge a chance to learn and grow.

So, with your radio in hand and frequencies at your fingertips, venture forth into the exciting world of amateur radio. The airwaves are waiting for you, filled with the voices of fellow enthusiasts eager to welcome another to the fold. The journey you're about to embark on is one of discovery, community, and endless possibilities. Welcome to amateur radio; your adventure starts now.

Appendix A:
Appendix

Welcome to the Appendix, a crucial cornerstone in charting your path through the fascinating world of radio communication. As we've journeyed from the basics of Baofeng radios to the broader landscape of ham radio, our adventure has been filled with discoveries, challenges, and triumphs. Think of this section not just as a summary but as your springboard into a world where knowledge meets passion, igniting endless possibilities.

Empowerment Through Knowledge

Embarking on the ham radio journey isn't merely about understanding frequencies or mastering equipment. It's about embracing empowerment. Each page you've turned, each concept you've grasped, brings you closer to becoming not just an enthusiast but a beacon within the community. The empowerment gained through this knowledge is your tool to explore, connect, and make a difference.

Practical Guidance for Real-World Adventures

The essence of this book lies in its practical guidance, designed to transition seamlessly into your radio adventures—be it for hobby, preparedness, or exploration. By now, you should feel equipped to navigate the layers of radio communication, apply best practices, and problem-solve with confidence. Remember, the journey is as rewarding as the destination, and every challenge is an opportunity for growth.

The Journey Ahead

As you stand on the threshold of your next adventure, remember that learning is a continuous journey. The world of ham radio is dynamic, evolving with advancements in technology and shifts in the global landscape. Stay curious, stay connected, and let your quest for knowledge be unending.

Your toolbox is now filled with the essentials: understanding of ham radio's history, its role in modern communication, the intricacies of Baofeng radios, and the fundamentals of antennas and equipment. You've also explored the practicalities of making contacts, the importance of operating procedures, and the pivotal role of amateur radio in emergency preparedness.

Charting Your Unique Path

What lies ahead is a path uniquely yours. With the foundation laid out in this book, you're now equipped to carve your journey in the amateur radio world. Explore further, connect deeper, and let your passion for radio communication light up the world. The amateur radio community is a tapestry of stories, innovations, and shared experiences—your story is waiting to be told.

In closing, I encourage you to revisit chapters, dive deeper into topics that pique your interest, and continuously seek out new knowledge. Join clubs, participate in forums, and attend hamfests. The world of ham radio is vibrant and welcoming, full of individuals eager to share their knowledge and experiences.

Remember, your journey in radio communication is both personal and collective. Embrace it with enthusiasm, curiosity, and a spirit of adventure. Welcome to the community—you're home.

As your adventures unfold, let this book be your guide, a companion in your exploration of the radio waves. The next chapter of your journey starts now. Let's make it unforgettable.

Glossary of Terms

As you embark on this adventure into the world of amateur radio and Baofeng radios, you'll soon discover that the journey is as rewarding as the destination. Let's demystify some of the terms you've encountered in this handbook, making the technical language of radio communication more accessible and less intimidating. Whether you're tinkering with your first Baofeng radio, preparing for your license exam, or venturing into outdoor adventures with a reliable communication tool, this glossary will be your compass.

Amateur Radio (Ham Radio)

Amateur Radio, affectionately known as *Ham Radio*, is a hobby and service where enthusiasts use various forms of radio communications equipment to connect with other radio amateurs for public service, recreation, and self-training.

Antenna

An essential component of radio equipment that is used to transmit and receive signals. Remember, the quality and type of antenna can significantly affect your radio's performance.

Baofeng

A popular brand of affordable and versatile handheld transceivers favored by amateur radio enthusiasts, preppers, and outdoor adventurers alike. Known for their ease of use and adaptability to various communication needs.

Call Sign

A unique identifier assigned to an amateur radio operator by their national telecommunications authority. It's your personal identifier on the airwaves, so wear it with pride.

FM, AM, SSB

- *FM (Frequency Modulation)*: A modulation technique used in radio broadcasting and for VHF and UHF communications where sound is encoded by varying the frequency of the wave.

- *AM (Amplitude Modulation)*: A modulation technique where the amplitude of the transmitted signal is made to vary in accordance with the information being sent.

- *SSB (Single Side Band)*: A form of amplitude modulation that uses less bandwidth and power, often used for long-distance communication.

Frequency Bands

Ranges of frequencies allocated for various purposes, including amateur radio. These bands are your playground, offering different experiences and opportunities to connect across the globe.

License Classes

In the US, there are three classes of amateur radio licenses: Technician, General, and Amateur Extra. Each class grants different privileges on frequency bands and modes of communication, enabling you to grow and explore new horizons.

Modulation Modes

Techniques used to convey information over radio waves. As you've discovered, there's more than one way to share a story, and each mode has its own charm and challenges.

Net Operations

Organized meetings on a specific frequency at a predetermined time, often led by a net control station. Think of it as a communal gathering, bringing together voices from near and far.

Repeater

A radio transmitter and receiver that automatically retransmits your signal at a higher power or on a different frequency, extending the range of communication. Like a lighthouse, it amplifies and guides your signal across vast distances.

Satellite Communications

Using satellites in orbit to communicate over long distances, transcending the limitations of line-of-sight communication. It's a leap into the future, connecting us in ways once imagined only in science fiction.

Embrace this journey into amateur radio with an open mind and a curious heart. Each term you learn, each concept you grasp, brings you closer to becoming not just an enthusiast, but a connoisseur of the airwaves. Let this glossary be your first step towards mastering the art and science of radio communication, illuminating your path to countless discoveries and connections. Remember, the airwaves are vast, and they whisper secrets to those who listen. So tune in, turn up the volume, and let the adventure begin.

Resources for Further Learning

Embarking on the journey of amateur radio is not merely about acquiring technical skills. It's a voyage of continuous learning, a door to a world where every piece of knowledge you gain further enriches your experience. This section aims to guide you through resources that will illuminate your path, broaden your horizons, and deepen your understanding of the fascinating world of Ham radio.

First and foremost, the *ARRL Handbook for Radio Communications* is an indispensable guide. Published annually by the American Radio Relay League, it's widely regarded as the bible for amateur radio enthusiasts. The handbook not only covers technical specifics and theory but also offers practical tips for both beginners and seasoned operators. It's a treasure trove of information that supports a hands-on approach to learning, helping you acquire the know-how to tackle various challenges in ham radio.

Diving into the practical side of things, YouTube provides an array of channels dedicated to amateur radio. Here, experienced hams demonstrate setup processes, review equipment, and share insights on overcoming operational hurdles. Channels like "Ham Radio Crash Course" and "QRZnow" are goldmines of information, presenting complex concepts in an easily digestible manner. The visual and interactive nature of these resources significantly enhances the learning experience, making the intricate world of radios accessible to all.

Online forums and communities such as *QRZ.com* and the *HamRadioForum* are vibrant platforms where enthusiasts gather to exchange knowledge, solve problems, and share their experiences. These communities are welcoming to newcomers, offering a wealth of collective knowledge. Engaging in these forums provides an opportunity to ask questions, seek advice, and connect with like-minded individuals who are more than willing to guide you through your amateur radio journey.

For a more structured approach to learning, consider enrolling in online courses or webinars. Platforms like *Coursera* and *Udemy* often host courses related to radio communication, electronic fundamentals, and signal processing. These courses range from beginner to advanced levels, catering to learners with varying degrees of expertise.

Books such as "Ham Radio For Dummies" by H. Ward Silver provide a gentle introduction to the world of amateur radio. Blending humor with expertise, Silver demystifies the subject, making it accessible to readers with no prior knowledge. This book serves as a fantastic starting point, offering clear explanations on getting started, getting licensed, and making your first contact.

For those interested in the technical intricacies and design aspects of radio equipment, "The ARRL Antenna Book" is a must-read. This comprehensive guide covers everything from basic antenna theory to advanced design, helping you to not only understand but also to craft antennas that enhance your radio experience.

Podcasts, such as "Ham Radio Workbench" and "The Ham Radio Show," offer insightful discussions on various aspects of amateur radio, from gear reviews to interviews with experienced hams. Listening to these podcasts can provide valuable insights and keep you updated on the latest trends and technologies in the ham radio world.

Participation in ham radio clubs and local groups is another fantastic way to learn and grow. Such communities often organize workshops, field days, and social events where you can gain hands-on experience under the guidance of veteran operators. These clubs serve as a support system, a place to form lifelong friendships, and a way to give back through volunteering and community service initiatives.

Exploring digital modes of communication opens up a new dimension in amateur radio. Resources like the "PSK31 Handbook" guide you through the nuances of this popular digital mode, helping you understand how to set up your station for digital transmission and receive. It's an exciting realm that combines traditional radio with the power of modern digital technology.

Software tools like *CHIRP* and *Ham Radio Deluxe* facilitate programming your radios and managing logs. Familiarizing yourself with these tools not only streamlines your operations but also introduces you to the broader ecosystem of amateur radio software. Tutorials and guides are readily available online, offering step-by-step instructions on leveraging these tools for a more efficient and enjoyable radio experience.

Documentaries and films on amateur radio not only entertain but also inspire. Watching stories of how ham radio has facilitated emergency communication during disasters, connected people across the globe, or even contacted the International Space Station can fuel your passion and appreciation for what amateur radio can achieve.

The *FCC's website* offers a plethora of official documentation and information regarding the legal aspects, frequencies, and regulations surrounding amateur radio. It's critical to familiarize yourself with these guidelines to ensure your activities remain within legal boundaries.

Lastly, don't overlook the value of experimenting and learning through direct experience. Building kits, tweaking settings, and simply playing around with your radio equipment can yield profound insights and a deep sense of satisfaction. It's through trial and error, successes, and failures, that you truly become proficient and innovative in amateur radio.

Remember, the world of amateur radio is vast and ever-evolving. The resources mentioned here are just the beginning. Stay curious, keep exploring, and never hesitate to reach out to the community for help and guidance. The journey of learning and discovery in amateur radio is as rewarding as the destination itself. Embrace each step, each challenge, and let your passion guide you to new heights.

FCC Rules and Regulations Summary

Embarking on the journey of amateur radio is not merely about acquiring a new hobby; it's entering a world rich with history, community, and regulation. The Federal Communications Commission (FCC) plays a pivotal role in this realm, setting forth rules that ensure the airwaves remain accessible, safe, and enjoyable for all. Understanding these regulations can feel daunting, but it's a crucial stepping stone toward becoming a responsible and successful ham radio operator.

At its core, the FCC's regulations serve to organize the use of radio frequencies so that they can be used efficiently and without interference. This is particularly important for amateur radio operators, as they share the airwaves with commercial, military, and other government entities. Each user has their own slice of the spectrum, and adherence to these allocations is non-negotiable.

A foundational aspect of FCC rules pertains to licensure. Everyone who wishes to operate on ham radio frequencies must pass an examination that confirms their understanding of both the technical aspects and legal responsibilities of radio operation. There are different classes of licenses, each granting privileges aligned with the operator's level of knowledge and expertise.

The FCC also places a strong emphasis on identification. Every amateur operator is assigned a unique call sign once they're licensed. This call sign serves as an audible signature on the airwaves, and regulations mandate its use at specific intervals during communication. It's not just a form of identification; it's a badge of honor and a symbol of one's commitment to the responsible use of the spectrum.

Operationally, the FCC sets forth rules on what can be communicated over the air. This includes prohibiting transmission of music, obscene content, and discussions intended for commercial gain. The guidelines seek to preserve the amateur radio bands as places for personal growth, technical experimentation, and emergency communication.

Interference is another major concern addressed by the FCC. Operators are urged to take all reasonable measures to avoid interfering with others' transmissions. This includes choosing the right frequencies, adhering to power limits, and employing equipment that minimizes unintended emissions. In a shared space, respect for others is paramount.

Speaking of equipment, the FCC has regulations on the types of devices and modifications that are permissible. These rules ensure that devices used by amateur operators are safe, both from an operational and a spectral integrity perspective. Modifications that might lead to interference or unsafe conditions are typically not allowed.

For those looking to push the boundaries of amateur radio, the FCC provides avenues for experimentation and innovation. Special licenses and provisions exist for those wishing to explore new technologies, modes of communication, or advancements in radio equipment. Herein lies the spirit of amateur radio: a blend of tradition and forward-thinking invention.

Emergencies bring another layer of FCC regulations into focus. Amateur radio operators are often called upon during disasters, and the FCC has rules that facilitate this critical role. These include allowing broader use of frequencies in times of emergency and guidelines for how amateurs can assist disaster relief efforts without interfering with official emergency services.

The FCC also recognizes the importance of international communication. Given that radio waves do not respect national borders, there are international agreements in place that the FCC incorporates into its regulations. This allows U.S. amateurs to communicate across the globe, fostering international goodwill and understanding.

It is worth noting that the FCC's rules are not static. They evolve in response to new technologies, changes in international regulations, and the needs of the amateur radio community. Keeping abreast of these changes is part of the ongoing commitment required from amateur operators. It's a dynamic field, and flexibility is key.

Enforcement of these regulations is another crucial function of the FCC. Violations can lead to warnings, fines, and even the revocation of one's license. The role of the FCC is not to punish, however, but to ensure that all operators are contributing to a healthy, productive radio environment. It's a reminder that with great privilege comes great responsibility.

As complex as these rules might seem at first, they are integral to maintaining the spirit and integrity of amateur radio. They are there to protect, to guide, and to ensure that this remarkable communication medium thrives. Remember, every seasoned operator started as a beginner, navigating these very regulations to find their voice on the airwaves.

Embrace the challenge. The rules and regulations set by the FCC are not hurdles but stepping stones towards becoming a proficient amateur radio enthusiast. With time, patience, and dedication, you will not only master these guidelines but also unlock the full potential that amateur radio offers. It's a journey that promises growth, community, and endless exploration.

Finally, becoming familiar with FCC rules and regulations is not just about compliance; it's about showing respect for the history and community of amateur radio. It's a pledge to uphold the values that have made this hobby a cornerstone of communication technology for over a century. As you embark on your amateur radio journey, let these principles guide you towards becoming not just an operator, but a steward of the airwaves.

www.ingramcontent.com/pod-product-compliance
Lightning Source LLC
Chambersburg PA
CBHW021143070326
40689CB00043B/1087